AN ATLAS OF WORLD POLITICAL FLASHPOINTS

To
BHPA
for all her diligence
and enthusiastic support

AN ATLAS OF WORLD POLITICAL FLASHPOINTS

A sourcebook of geopolitical crisis

EWAN W. ANDERSON

University of Durham

Maps produced by **Gareth Owen**
City Cartographic
under the direction of
Don Shewan
City of London University

Pinter Reference
LONDON

Pinter Publishers
25 Floral Street, Covent Garden, London, WC2E 9DS, United Kingdom

First published in 1993

© Text, Ewan W. Anderson; Maps, Pinter Publishers

Apart from any fair dealing for the purposes of research or private study, or criticism or review, as permitted under the Copyright, Designs and Patents Act, 1988, this publication may not be reproduced, stored or transmitted, in any form or by any means, or process without the prior permission in writing of the copyright holders or their agents. Except for reproduction in accordance with the terms of licences issued by the Copyright Licensing Agency, photocopying of whole or part of this publication without the prior written permission of the copyright holders or their agents in single or multiple copies whether for gain or not is illegal and expressly forbidden. Please direct all enquiries concerning copyright to the Publishers at the address above.

Ewan Anderson is hereby identified as the author of this work as provided under Section 77 of the Copyright, Designs and Patents Act, 1988.

British Library Cataloguing in Publication Data
A CIP catalogue record for this book is available from the British Library

ISBN 1 85567 053 4

Typeset by BookEns Ltd., Baldock, Herts.
Printed and bound in Great Britain by SRP Ltd., Exeter.

Contents

Preface	vii	
List of Acronyms	viii	
Key to Maps	ix	
World Map	x	
Introduction	xiii	

#	Entry	Page
1.	Abu Musa	1
2.	The Aegean Sea	5
3.	The Åland Islands	9
4.	The Alps-Adriatic region (the former Yugoslavia)	11
5.	Antarctica	13
6.	The Aozou Strip	17
7.	The Attila Line	19
8.	Bab el Mandeb	23
9.	The Baltic Republics	25
10.	The Barents Sea	29
11.	The Basque Country/Euzkadi	31
12.	The Beagle Channel	35
13.	The Benguela Railway	37
14.	Berlin	41
15.	Bessarabia	43
16.	Cabinda	47
17.	The Caprivi Strip	49
18.	Ceuta	51
19.	The Chagos Archipelago (Diego Garcia)	53
20.	The Curzon Line	57
21.	East Timor	59
22.	Epirus	61
23.	Eritrea	65
24.	The Falkland Islands (Malvinas)	67
25.	The Gaza Strip, the Golan Heights and the West Bank	71
26.	Guantanamo	77
27.	Guyana	79
28.	The Hatay	83
29.	The Hawar Islands	85
30.	Hong Kong	89
31.	The Strait of Hormuz	91
32.	Jan Mayen Island	95
33.	Karelia	97
34.	Kashmir	101
35.	The Kola Peninsula	103
36.	Kosovo	107
37.	Kurdistan	109
38.	The Kurile Islands	113
39.	The Liancourt Rocks	117
40.	Macao	119
41.	The McMahon Line	121
42.	The Magellan Strait	123
43.	The Strait of Malacca	127
44.	Mayotte Island	129
45.	Mururoa Atoll	133
46.	Nagorno-Karabakh	135
47.	Navassa Island	139
48.	The Neutral Zones	141
49.	Northern Ireland	143
50.	Lake Nyasa (Malawi)	149
51.	The Oder–Neisse Line	153
52.	The Ogaden	155
53.	The Panama Canal	157
54.	The Paracel Islands	161
55.	The Rann of Kutch	163
56.	Rockall	165
57.	The Sahel	167
58.	Senegambia	171
59.	The Senkaku and Ryukyu Islands	173
60.	The Shatt al Arab	177
61.	The Sinai Peninsula and Taba	179
62.	The Sino-Russian (formerly Soviet) Border	183
63.	South Lebanon	187
64.	Spitzbergen (Svalbard)	191
65.	The Spratly Islands	193
66.	The Suez Canal	197
67.	Surinam	199
68.	Tacna	203
69.	The Tanzam Railway	205
70.	The Strait of Tiran	209
71.	The Gulf of Tongking	211
72.	Transylvania	215
73.	Trieste	217

74.	The Tromelin Island	221
75.	The Tunbs Islands	223
76.	The Tyrol	225
77.	The Wakhan Panhandle	227
78.	Walvis Bay	231
79.	Warbah and Bubiyan Islands	233
80.	Western Sahara	237

Index 239

Preface

Since late 1989, the global geopolitical scene has changed so profoundly that the framework for security accepted throughout the Cold War period has been shattered. The reduction in East–West tension has not only necessitated a completely new assessment of potential flashpoints but has allowed a reappraisal of past events.

In this atlas major areas of conflict or potential conflict are examined, and a few examples of historical significance are included. For each the geographical situation, historical background, relative importance and current status are considered. Each account is illustrated with a map of the immediate region of the flashpoint, specially compiled and drawn for this Atlas by City Cartographic at the City of London Polytechnic, under the direction of Don Shewan.

Clearly in a period of such geopolitical volatility it is not possible to forecast all areas of possible conflict. However, most of the more generally recognized flashpoints have been included. This has involved a very detailed data search, and for this I am most grateful to my student and colleague Clive Schofield. I must also thank Bid Austin for her careful typing and editing and Barry Austin for his meticulous checking of the manuscript. For any errors, the responsibility is mine.

Ewan Anderson
Durham
July 1992

List of Acronyms

ASEAN	Association of South East Asian Nations	RMDSZ	Democratic Association of Hungarians in Romania
BIOT	British Indian Ocean Territory	SAC	Supreme Allied Commander
bpd	barrels per day	SADF	South African Defence Forces
CIS	Commonwealth of Independent States	SADR	Sahrawi Arab Democratic Republic
dwt	dead-weight tonnage	SAS	Special Air Service
EC	European Community	SLA	South Lebanon Army
EEZ	Exclusive Economic Zone	SPC	South Pacific Commission
ENI	Ente Nationale Idrocarburi	SSR	Soviet Socialist Republic
EOKA	National Organization of Freedom Fighters	SUMED	Suez-Mediterranean
		SVP	Sudtiroler Volks Partie
EPLF	Eritrean People's Liberation Front	SWAPO	South West Africa People's Organization
ETA	Euzkadi ta Askatasuna (Basque Homeland and Freedom)	UAE	United Arab Emirates
FNLA	National Front for the Liberation of Angola	UDA	Ulster Defence Association
		UK	United Kingdom
FRETELIN	Revolutionary Front for Independence (East Timor)	ULCC	Ultra-Large Cargo Carrier
		UN	United Nations
GCC	Gulf Cooperation Council	UNCLOS	United Nations Conference on the Law of the Sea
GDP	Gross Domestic Product		
GMR	Great Manmade River	UNEF	United Nations Emergency Force
HEP	Hydro-Electric Power	UNFICYP	United Nations Peace-keeping Force in Cyprus
IJC	International Court of Justice		
IRA	Irish Republican Army	UNHCR	United Nations High Commission for Refugees
MINURSO	Mission for the Referendum In Western Sahara		
		UNIFIL	United Nations Interim Force in Lebanon
MPLA	Popular Movement for the Liberation of Angola		
		UNITA	National Union for the Independence of Angola
MPM	Mouvement Populaire Mahorais		
MSI	Movimento Sociale Italiano	US	United States
NATO	North Atlantic Treaty Organization	USSR	Union of Soviet Socialist Republics
		VLCC	Very-Large Cargo Carrier
nml	nautical miles	WMO	World Meterological Organization
OAS	Organization of American States	ZOPFAN	Zone of Peace, Freedom and Neutrality
OAU	Organization of African Unity		
PLO	Palestine Liberation Organization		
PNG	Papua New Guinea		
PNV	Partido Nacionalista Vasco		

Key to Maps

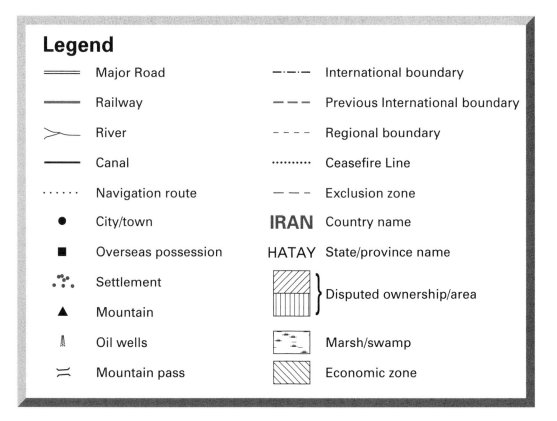

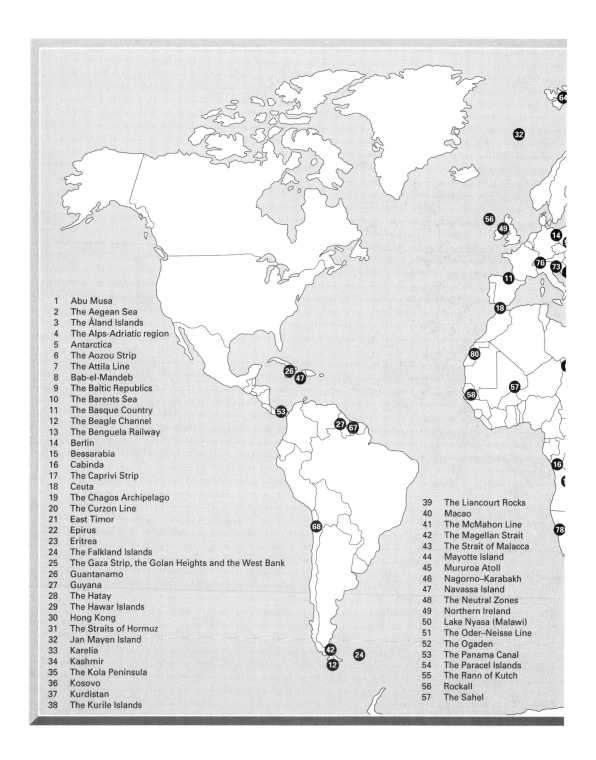

1	Abu Musa
2	The Aegean Sea
3	The Åland Islands
4	The Alps-Adriatic region
5	Antarctica
6	The Aozou Strip
7	The Attila Line
8	Bab-el-Mandeb
9	The Baltic Republics
10	The Barents Sea
11	The Basque Country
12	The Beagle Channel
13	The Benguela Railway
14	Berlin
15	Bessarabia
16	Cabinda
17	The Caprivi Strip
18	Ceuta
19	The Chagos Archipelago
20	The Curzon Line
21	East Timor
22	Epirus
23	Eritrea
24	The Falkland Islands
25	The Gaza Strip, the Golan Heights and the West Bank
26	Guantanamo
27	Guyana
28	The Hatay
29	The Hawar Islands
30	Hong Kong
31	The Straits of Hormuz
32	Jan Mayen Island
33	Karelia
34	Kashmir
35	The Kola Peninsula
36	Kosovo
37	Kurdistan
38	The Kurile Islands
39	The Liancourt Rocks
40	Macao
41	The McMahon Line
42	The Magellan Strait
43	The Strait of Malacca
44	Mayotte Island
45	Mururoa Atoll
46	Nagorno–Karabakh
47	Navassa Island
48	The Neutral Zones
49	Northern Ireland
50	Lake Nyasa (Malawi)
51	The Oder–Neisse Line
52	The Ogaden
53	The Panama Canal
54	The Paracel Islands
55	The Rann of Kutch
56	Rockall
57	The Sahel

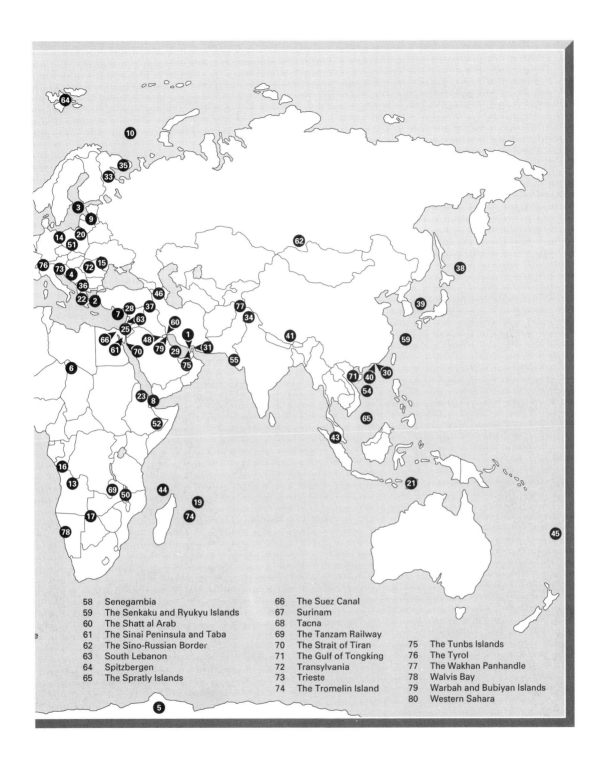

58	Senegambia
59	The Senkaku and Ryukyu Islands
60	The Shatt al Arab
61	The Sinai Peninsula and Taba
62	The Sino-Russian Border
63	South Lebanon
64	Spitzbergen
65	The Spratly Islands
66	The Suez Canal
67	Surinam
68	Tacna
69	The Tanzam Railway
70	The Strait of Tiran
71	The Gulf of Tongking
72	Transylvania
73	Trieste
74	The Tromelin Island
75	The Tunbs Islands
76	The Tyrol
77	The Wakhan Panhandle
78	Walvis Bay
79	Warbah and Bubiyan Islands
80	Western Sahara

Introduction

The epithet 'geopolitical', used in the subtitle, encapsulates the focus of this atlas. The issues depicted and discussed all involve the interplay of geography and politics. They are geopolitical in that the concern is with the geographical setting in which political decisions are taken or, at worst, conflict occurs. The emphasis is upon those geographical factors with the potential to influence such political events.

Since the first use of the term by Kjellen (1917), the concept of geopolitics has changed considerably (Agnew and Anderson 1992). The approach of diplomats and strategists from Mackinder (1904) to Walters (1974) has depended largely upon a consideration of the influence of physical geography upon politics at the international level. More recently, Cohen (1973) and Child (1985), in particular, have taken a far broader view of the contribution of geography, while, in his extremely catholic use of the term, Kissinger (1982), clearly depends on inputs from most facets of the subject. The obvious extension from the physical and often strategic aspects has been into economic geography and terms such as 'resource geopolitics' are now commonly employed. It should perhaps be stressed that this development has been one of interpretation, since Kjellen made specific provision in his terminology for resources as a key geographical variable. Social geography, broadly defined to include, for example, cultural, ethnic and linguistic variations, also offers an important contribution to geopolitics. Furthermore, many of the issues are strongly influenced by the historical dimension and historical geography is therefore of significance.

The geopolitical crises and potential crises discussed in the atlas, in many cases present a complex mosaic of geographical contributions. However, there are examples in which each of the aspects of the subject identified, predominates. Karelia, the Åland Islands, the Sahel and the Sinai Peninsula are all basically concerns of physical geography. Whereas, in the case of the Gaza Strip, the Golan Heights, the West Bank, the Chagos Archipelago, Ceuta and the Kola Peninsula, the primary considerations are strategic. Problems or potential problems over Bab el Mandeb, the Benguela Railway, the Hawar Islands, the Neutral Zones, the Paracel Islands, Rockall and the Shatt al Arab are dominated by economic geography. In contrast, while there are both physical and economic components, the Attila Line, the Basque country, Bessarabia, East Timor, Epirus, Kosovo, Kurdistan, Northern Ireland and Transylvania are primarily flashpoints as a result of social geography. Historical geography is important in many of the issues, such as those of the Curzon Line, the Falkland Islands, Guantanamo, the Hatay, the Kurile Islands, Macao, the Sino-Russian Border and Tacna and Trieste, while Berlin must be considered a geopolitical flashpoint only in historical terms.

In purely spatial terms, the geographical input to geopolitics can be envisaged in terms of points, lines and areas. Very small areas, essentially points, can be the epicentres of geopolitical upheaval, with wide ranging consequences. Such points would include: Abu Musa, Guantanamo, the Liancourt Rocks, Mururoa Atoll, the Spratly Islands and the Tunbs Islands. The main linear features are boundaries and lines of communication. Boundaries are a major concern in the case of Antarctica, the Aozou Strip, the Barents Sea, the West Bank, the Hawar Islands, Kashmir, the Kurile Islands, the McMahon Line, the Oder-Neisse Line and the Rann of Kutch. Potential crises over communications include: Bab el Mandeb, the Benguela Railway, the Strait of Hormuz, the Strait of Malacca, the Panama Canal, the Shatt al Arab, the Suez Canal, the Tanzam Railway and, in a totally different context, the Caprivi Strip and the Wakhan Panhandle. Among the areal or territorial concerns are: the Alps-Adriatic region, the Falkland Islands, Guyana, Karelia, Kurdistan, Nagorno-Karabakh, the Ogaden, South Lebanon, Transylvania, the Tyrol and Western Sahara. In most of these cases there is some overlap in that, for example, concerns over territory must, at some stage, involve boundary considerations, but such a geographical classification provides a further focus for anaylsis.

While views on the geographical component of geopolitics have changed, the political has remained focused on one scale only, the global level. Although it has obviously been realized that, for example, the internal politics of Iran might affect the security of the Strait of Hormuz, the contributions of national and regional political systems have not been formalized. Although political decision-making at the national and regional level is likely to be predominantly of interest at that level only, some results may have far wider and even global significance. The problems of the Basque Country, Bessarabia, Eritrea, the Kola Peninsula, Northern Ireland and Walvis Bay are essentially national or even local in nature. Regional crises or potential crises include: the Aegean Sea, the Benguela Railway, the Chagos Archipelago, East Timor, Epirus, Hong Kong, Kashmir, Nagorno-Karabakh, the Neutral Zones, the Rann of Kutch, Senegambia, Spitzbergen, Western Sahara. As a result of local, national or regional political actions, any of these would become global geopolitical flashpoints. In contrast, as a result of their international nature or importance, the following are already global concerns: Antarctica, Bab el Mandeb, the Gaza Strip, the Golan Heights, the West Bank, the Strait of Hormuz, Kurdistan, the Strait of Malacca, Mururoa Atoll, the Panama Canal, the Shatt al Arab and the Suez Canal.

Thus, the range of issues addressed in the atlas, selected because of their geopolitical significance, includes problems which illustrate the breadth of the geographical and the political components.

References

Agnew, C.T. and Anderson E.W. (1992), *Water Resources in the Arid Realm*, Routledge, London.

Child, J. (1985), *Geopolitics and Conflict in South America*, Praeger, New York.

Cohen, S. (1973), *Geography and Politics in a World Divided*, Oxford University Press, New York.

Kissinger H. (1982), *American Foreign Policy: A Global View*, Institute of Southeast Asian Studies, Singapore.

Kjellen, R. (1917), *Der Staat als Lebensform*, K. Vowinckel, Berlin.

Mackinder, H.J. (1904), 'The Geographical Pivot of History', *Geographical Journal*, 23, 423–37.

Walters, R.E. (1974), *The Nuclear Trap, An Escape Route*, Penguin, Harmondsworth.

1 Abu Musa

Description

Abu Musa, approximately 5 km across and with good deep-water anchorages, lies in a strategic position at the entrance to the Persian/Arabian Gulf, some 69 km from Iran and 56 km from Sharjah (United Arab Emirates). The latest available figure for the population is approximately 800. If an Exclusive Economic Zone (EEZ) were constructed around Abu Musa and the Tunbs Islands, using median lines, its area would be some 1,500 square nautical miles (nml).

History and importance

The sovereignty over Abu Musa and the Tunbs Islands has been disputed between Iran (formerly Persia) and the United Arab Emirates (UAE), initially the Emirates of Sharjah and Ras al Khaymah over a long period. The Qawasim family of the then Trucial Coast assumed possession of the three islands by the mid-19th century and this ownership was recognized by Britain. However, if a median line were drawn down the Gulf, only Abu Musa would be considered Arab. Persia continued to press its claims and in 1904 introduced into the islands customs officials, who were only withdrawn following British protests. A little later, Britain forcibly ejected workers of a German company who were mining iron oxide on Abu Musa.

In 1913, Britain re-emphasized its support for the Arab claim by erecting a lighthouse on Greater Tunb. However, in 1921, a split in the Qawasim family resulted in the separation of the two Emirates of Sharjah and Ras al Khaymah, with Abu Musa being regarded as belonging to the former and the Tunbs Islands to the latter. The dispute flared again in 1928 when Persian customs officials seized an Arab dhow at Greater Tunb. Despite a verbal agreement between London and Tehran in that year to accept the status quo (that is Persian sovereignty over Sirri, a small island in the Persian/Arabian Gulf, in the approaches to the Strait of Hormuz; and Arab sovereignty over Abu Musa and the Tunbs Islands), Persian claims were restated. Since the Tunbs Islands are on the Iranian side of the median line, their status was at least negotiable in theory, but Ras al Khaymah refused to consider the issue of sovereignty or even an Iranian lease.

The key factors which revived the issue of sovereignty in the 1960s were concerns over oil transport through the Strait of Hormuz and Iranian fears of a power vacuum in the wake of the British withdrawal from the Gulf which was announced in 1968. In mid-November 1971, just before the British withdrawal, agreement was reached between Iran and Sharjah whereby Iran would take control over the strategic areas of Abu Musa in return for a payment to Sharjah of £0.5 million annually for nine years or until Sharjah's oil revenues reached £3 million per annum.

Thus, on 30 November 1971, two days before the official British withdrawal from the Gulf and the independence of the UAE, Iranian troops occupied Abu Musa. They also laid claim to the Tunbs Islands, which were only captured after a bloody encounter with a small Ras al Kahymah–British detachment, an action that caused a storm of criticism of both Iran and Britain throughout the Arab world.

Indeed, during the Iran–Iraq War, Iraq, in attempting to consolidate Pan-Arab support, announced its intention of restoring the islands to Arab control. In fact, during the early stages of the war, Iraq was only restrained by Saudi Arabia from bombing Iranian positions on the islands from either the UAE or Oman. In 1982, the Iranians threatened to block the Strait of Hormuz, possibly using the islands, but exactly how this was to be achieved was never revealed.

Status

The islands remain firmly under Iranian control. Because of their position at the entrance to the

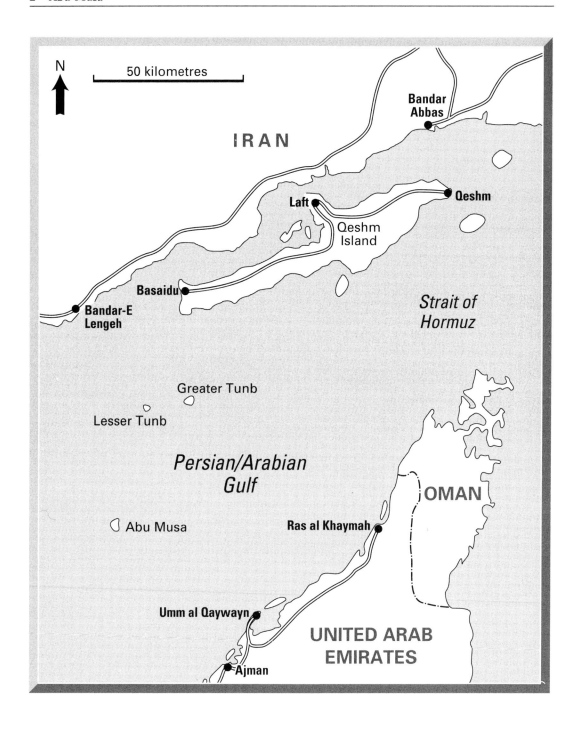

Strait of Hormuz, Iran considers them vital to its oil security. The bulk of Iranian oil exports still pass through the Gulf, and even in the time of the Shah it was considered Iran's jugular. For similar reasons, the Arab states also view the position of the islands as highly strategic. In their case, potential problems for oil exports are allied to the fear of a fundamentalist Iran obtaining a stranglehold on the Gulf through military outposts on the islands.

The 1971 agreement between Sharjah and Iran left the status of Abu Musa both ambiguous and ill-defined, since both countries claim full sovereignty. Iran declined to demarcate the 4 km-long boundary across the island, separating the Iranian held strategic points in the north from the southern part of the island, including the village of Abu Musa itself, controlled by Sharjah. This dispute remains a major irritant in Arab–Iranian relations in the Gulf.

References

Blake, G.H. and Drysdale, A. (1985), *The Middle East and North Africa: A Political Geography*, Oxford University Press, Oxford.

Day, A.J. (ed.) (1984), *Border and Territorial Disputes*, Longman, London.

Hiro, D. (1985), *Iran Under the Ayatollahs*, Routledge & Kegan Paul, London.

Peterson, J.E. (1985), 'The Islands of Arabia: Their Recent History and Strategic Importance', *Arabian Studies*, **VII**, pp. 23–35.

Prescott, J.R.V. (1985), *The Maritime Political Boundaries of the World*, Methuen, London.

Rais, R.B. (1986), *The Indian Ocean and the Superpowers*, Croom Helm, London.

Swearingen, W.D. (1981), 'Sources of Conflict over Oil in the Persian/Arabian Gulf', *The Middle East Journal*, 35, No. 3, Summer, Middle East Institute, Washington DC, pp. 314–30.

2 The Aegean Sea

Description

The most easterly of the semi-enclosed seas on the northern side of the Mediterranean, the Aegean Sea is dominated by islands, almost all Greek-owned. As with many of the marginal seas of the Mediterranean, there is potential for mineral extraction from the sea, while, strategically, it guards the entry to the Turkish Straits. Viewed on a larger scale, it lies between Greece and Turkey, two NATO allies who have had a number of disputes, most obviously illustrated by the division of Cyprus. Globally, with the demise of the Soviet Union, the importance of the Black Sea Fleet may have diminished, thereby reducing the status of the Turkish Straits.

History and importance

Since its independence from Turkey in 1830, Greece has progressively established sovereignty over the islands of the Aegean, benefiting particularly from the defeat of Turkey in 1913 in World War I and of Italy in 1943 in World War II. The result is that today Turkey has ownership of only two major islands: Gokceada (Imbros) and Bozcaada (Tenedos), both in the far north.

While neither side officially disputes this sovereignty, the result is that a string of Greek islands commands the Anatolian coastline. If the median line between these islands and the coast of Turkey were used as a boundary, the Aegean Sea would effectively be Greek-owned. The Turkish case is that the boundary should be a median line between the two mainlands. This claim is rejected by Greece, which cites the 1958 Convention on the Continental Shelf, which allows islands full effect (they can claim full offshore boundaries but some small islands e.g. Lampedusa claim only half effect). Greece has pressed for an International Court of Justice (ICJ) judgment, but the Court has declined to become involved in the case.

In support of its case, Turkey cites the judgment over the United Kingdom Channel Islands and the concessions made by Australia to Papua New Guinea (PNG) in the Torres Strait where Australian islands lie close to the PNG coastline. Additionally, both sides claim territorial waters extending for 6 miles, and Greece has indicated its desire to increase these to 12 nml. However, Turkey has made it clear that such an action would constitute a *causus belli* as the Greek share of the Aegean would then rise from 35 per cent to 66 per cent, whereas Turkey's would only increase from 9 to 10 per cent. Furthermore, Turkish vessels transiting from its Aegean ports would have to pass through Greek territorial waters.

Status

In East–West terms, the importance of the Aegean Sea has declined, but, in Greek–Turkish relations it has become, if anything, even more of a flashpoint. Current disputes have been exacerbated by threats from Greece to militarize the Aegean islands and there are also disputes over air space. Ironically, the prospects of discovering significant oil deposits appear modest, and only one field, producing 25,000 barrels per day (bpd), located near Thasos, is in production.

Given the intransigence on both sides, the emotive nature of the issues involved, the mutual suspicion and hostility and the recurrent problem of Cyprus (see Chapter 7, Attila Line), there seems little prospect of a peaceful solution in the medium term.

References

Beeley, B.W. (1989), 'The Turkish–Greek Boundary: Change and Continuity' in *International Boundaries and Boundary Conflict Resolution*, 1989 Conference Proceedings, International Boundaries Research Unit, University of Durham, pp. 29–40.

Blake, G.H. and Drysdale, A. (1985), *The Middle East and*

North Africa: A Political Geography, Oxford University Press, Oxford.
Day, A.J. (ed.) (1984), *Border and Territorial Disputes*, Longman, London.
Downing, D. (1980), *An Atlas of Territorial and Border Disputes*, New English Library, London.
The Economist (1991) 2 March.
Foreign Policy Institute (Ankara), 'Views on the Questions Between Turkey and Greece' in *Dis Politika* (Foreign Policy), **X**, Nos 3 and 4.
Munir, M. (1976), 'The Aegean Conflict: Is Reconciliation Possible?' *The Middle East*, No. 24, October, pp. 8–12.
Prescott, J.R.V. (1985), *The Maritime Political Boundaries of the World*, Methuen, London.
Rozakis, C.L. (1975), *The Greek-Turkish Dispute over the Aegean Continental Shelf*, Occasional Paper No. 27, Law of the Sea Institute, University of Rhode Island, Kingston, Rhode Island.

3 The Åland Islands

Description

The Åland Islands, with a population of some 23,000, constitute strategically important Finnish territory lying centrally at the entrance to the Gulf of Bothnia. Under Swedish rule until 1809, they were seized by Russia and joined to Finland. In 1854, the British and French fleets destroyed the fortifications of the islands, which, on the ending of the Crimean War in 1856, were demilitarized and neutralized but remained attached to Finland.

History and importance

On 15 November 1917, Finland declared independence from the Russian Empire, but in the previous August, the islanders had informed the King of Sweden of their desire to reunite with Sweden. Two plebiscites confirmed this decision, which was supported by the King of Sweden, and in February 1920 a Swedish military expedition secured the islands.

Having rejected a Finnish law granting autonomy within the Finnish state, in 1920 the council of the islands, together with the British government, called on the League of Nations to deal with the issue. The League recommended that the islands remain part of Finland, but with substantial autonomy, guarantees, demilitarization and neutralization. This status was accepted at the London Convention of 20 October 1921, signed by all the Baltic States, together with Britain, France and Italy.

However, with the threat of war in 1939 between Finland and the Soviet Union, Finland and Sweden agreed on 8 January to limited remilitarization and fortification of the islands. This was strongly rejected by the islanders, who presented a petition to the League of Nations, and Sweden subsequently withdrew from the proposals on 3 June. The Finnish decision to press ahead with fortification was finally reversed after the Finno–Soviet Peace Treaty, ratified 15 March 1940, when Finland proceeded to demilitarize the islands once more.

At the end of World War II on 12 September 1945, the Landsting (Parliament) of the Åland Islands demanded reunion with Sweden, but the Peace Treaty between the Allies and Sweden, signed 10 February 1947, stated that the islands would remain demilitarized as part of Finland.

Status

According to the October 1921 Treaty:

> In the eventuality of the Åland Islands, or through them, the mainland of Finland becoming the object of a sudden attack endangering the neutrality of the zone, Finland is to take necessary measures in the zone to stop or repel the aggression.

After World War II, the islands remained important primarily because the channels between them and the Finnish mainland constituted a transit route for Soviet submarines beyond Swedish surveillance. The dispute between Sweden and Finland has long since subsided, and the agreed maritime boundary signed on 15 January 1975 has been demarcated with the Åland Islands on the Finnish side. With the demise of East–West confrontation, the strategic significance of the islands has apparently declined, but, in a volatile situation with new states emerging in the Baltic, the issues could again be renewed.

References

Cartactual (1975), CA 53/13, 11, No. 53.
Day, A.J. (ed.) (1984), *Border and Territorial Disputes*, Longman, London.
Prescott, J.R.V. (1985), *The Maritime Political Boundaries of the World*, Methuen, London.
United States Department of State (1976), *Continental Shelf Boundary: Finland–Sweden*, Limits in the Seas, No. 71, 16 June, Bureau of Intelligence and Research, Washington DC.

10 The Alps-Adriatic region (the former Yugoslavia)

4 The Alps-Adriatic region (the former Yugoslavia)

Table 4.1 Proportions of different nationalities in each republic in Yugoslavia (%)

	Muslims	Croats	Macedonians	Montenegrins	Serbians	Slovenes	Albanians	Hungarians	Other
Bosnia	43.7	17.5			31.5				7.3
Croatia		70			11				19
Macedonia	2.1		67		2.3		19.7		8.9
Montenegro	13.4			68	6.5		5		7.1
Serbia	2				85.4				12.6
Slovenia		3				90			7
Kosovo					13.2		85		1.8
Vojvodina		5.4			54.4			19	21.2
YUGOSLAVIA	8.9	19.8	6	2.6	36.3	7.8	7.7	1.9	9

Source: 1981 census

Description

Yugoslavia, the land of the southern Slavs, was the most ethnically diverse country in Europe and was welded together as a result of shared privation in its recent history and the will of one man, President Josef Tito. He formed it into a complex multi-national federation with a three-tier system of national grouping. In the event, no single group had an overall majority in the federation and each republic had large minority groups (see Table 4.1). Apart from the local ethnic groups, there were also significant numbers of Muslims, Albanians and Hungarians.

History and importance

After World War II, the communist rulers of Yugoslavia had close ties with the Soviet Union, but these were broken in 1948. This left Yugoslavia isolated as the one major Communist state in Europe outside the Warsaw Pact. When, in 1989, the Warsaw Pact itself began to crumble, this eliminated much of the pressure on Yugoslavia and removed the key threat to national stability. At the time of writing, the country is breaking up and there is a possibility that, with continuing conflict, it will 'Balkanize' totally.

In the 17th and 18th centuries, there was active encouragement from the Austro-Hungarian Empire for military settlement in Slovenia and Croatia, to provide a buffer zone against Ottoman expansion. Many Serbs, fleeing from Ottoman rule, also settled in the region. Thus, Slovenia and Croatia have long ties with Austria and Hungary; and Croatia, in particular, has a sizeable Serbian minority.

The Austro-Hungarian military frontier was abolished in 1881 and by the census of 1910 Serbs constituted 24.6 per cent and Croats 62.5 per cent of the combined population of Slovenia/Croatia. Meanwhile, there was a demand for a Greater Croatia, to include Croatia itself, Slovenia and Bosnia-Hercegovina and to enjoy the same status as Austria and Hungary. In the area demanded there was no official place for Serbs. However, in 1908 Bosnia was annexed, following a growing, if short-lived, Serbian-Croat-South Slav nationalism and reconciliation generated in the face of Austrian oppression.

Yugoslavia finally became independent in 1919, but Croat nationalism forced the government to allow increased Croatian autonomy in the 1930s. Prior to the present, the climax of Serb-Croat rivalry was reached in a bitter and bloody civil war, fought during World War II. The Croat state which existed from 1941–5 included Bosnia and Hercegovina, together with northern Serbia. Serbia gained complete independence from provisions of the Treaty of Berlin in 1878.

However, after World War II, the borders of the republics reverted to those of the Austro-Hungarian and Ottoman Empires. They were ratified in 1946, with Croatia retaining thereby a substantial Serb minority. Conflict was suppressed by the authoritarian nature of the Communist rule. The 1974 Constitution recognized six republics and two autonomous regions (Kosovo and Vojvodina, now both incorporated into Serbia) and, apart from the Slovenia–Croatian border, the boundaries delimited in 1946 did not follow ethnic divisions.

Status

The region is in a continuing state of crisis, basically as a result of conflicts over various minority groups, particularly those in enclaves along the Bosnian and eastern Slovenian borders. Fighting erupted in August 1990 and again in March and April 1991. Since then, the conflict has spread and intensified at a bewildering rate. By July 1992, international concern over the plight of the citizens of the beleaguered Bosnian capital Sarajevo made intervention by Western military forces a possibility to back-up the United Nations peace-keeping forces already in place. At the time of writing the likely scenario of future events is unclear.

The present crisis was precipitated by Slovenian and Croatian unilateral declarations of independence on 26 June 1991. As a result the Federal Army of Yugoslavia, aided by Serbian irregulars, advanced into Croatia, capturing most Serbian-dominated areas and occupying Bosnia. At the same time, the Serbian minority population had declared the 'Serbian Autonomous Region of Karajina', centred on the Bosnian border enclaves. On 15 October 1991 Bosnia declared itself a sovereign state, and the European Community (EC) stated that any of the former Yugoslav republics could apply for recognition as independent states from 15 January 1992. Germany preempted the issue by stating that it would recognize both Slovenia and Croatia on that date. Meanwhile, Serbia has established an élite strike force as a prelude to forming its own republican armed forces and lasting peace seems, in the medium term, unlikely. Unless there is a good prospect of peace, peace-keepers from the EC or the United Nations are unlikely to take up permanent positions.

On the diplomatic front, the EC foreign ministers, meeting in Brussels on 17 December 1991, invited all six Yugoslav republics to apply for recognition as independent states, provided they fulfilled a set of stringent requirements. Meanwhile, despite the diplomatic victories of Croatia, Serbia appears to be winning the military conflict.

References

Boundary Bulletin, No. 1, (1990), International Boundaries Research Unit, Durham University.
Boundary Bulletin, No. 2, (1991), International Boundaries Research Unit, Durham University.
The Economist (1991) 25 May.
The Economist (1991) 6 June.
The Economist (1991) 7 July.
The Economist (1991) 3 August.
The Economist (1991) 10 August.
The Economist (1991) 31 August.
The Economist (1991) 8 September.
The Economist (1991) 21 September.
The Economist (1991) 5 October.
The Economist (1991) 12 October.
The Economist (1991) 23 November.
Englefield, G. (1991), 'Dividing Yugoslavia', *Boundary Bulletin*, No. 2, International Boundaries Research Unit, Durham University, pp. 9–12.
Muir, R. (1981), *Modern Political Geography*, Macmillan, London.
Sallnow, J. (1989), 'Yugoslavia – Powder Keg of Europe', *Geographical Magazine*, May, pp. 16–20.

5 Antarctica

Description

The great southern land mass on the globe, the Antarctic mainland, occupies an area of 13.9 million km^2, representing some 21.5 per cent of the earth's land surface. It is therefore larger than all of Africa, but only approximately 2 per cent of its surface is free of ice. Indeed, the covering of Antarctica represents 90 per cent of the world's ice, or 75 per cent of its freshwater store. If a complete melting occurred, the oceans would rise by approximately 60 m.

Unlike the Arctic, separating the United States and the former Soviet Union, it was never claimed that Antarctica had any major immediate strategic significance. International conflict has been between those supporting the conservation of the environment and those looking for economic development, especially of mineral resources. Most recently, the discovery of a hole in the ozone layer over Antarctica has brought prominence to the region.

History and importance

For a variety of reasons – economic, scientific and pseudo-strategic – Antarctica has achieved an increasing significance this century. The first formal claim was made by Britain in 1908, but parts of this were later ceded to Australia and New Zealand. Interest revived immediately prior to World War II: Hitler dispatched a sea-plane to Antarctica in 1938; as a result of this action, France and Norway registered claims in 1939. Argentina and Chile, basing their arguments on early exploration and the Treaty of Tordesillas of 1493, laid their claims in 1940 and 1941, respectively.

After World War II, it was thought that Antarctica might have some military significance and a network of Soviet installations was developed. The United States' navy mounted the largest ever expedition, comprising 4,000 men and 13 ships, including an aircraft carrier. By 1948, tension focused on the overlapping claims of Argentina, Chile and Britain. There was even one shooting incident between Argentinian and British troops on the Antarctic ice.

However, the International Geophysical Year (1957–8) precipitated the Antarctic Treaty which was signed on 1 December 1959 and entered into force on 23 June 1961, after ratification by 12 signatories: Argentina, Australia, Belgium, Chile, France, Britain, Japan, New Zealand, Norway, South Africa, the Soviet Union and the United States. Later, Brazil, Bulgaria, Denmark, Czechoslovakia, East Germany, West Germany, India, the Netherlands, Poland, Romania and Uruguay acceded to the Treaty organization. After conducting substantial scientific research, Poland and West Germany acquired consultative status.

Nevertheless, the legal status of the various claims remains in doubt. Some nations have staked traditional sovereignty claims, some have reserved the right to do so, while others believe that Antarctica should be managed on a consortium basis or the administration should be turned over to the United Nations and the continent become a 'World Park'. Several Antarctic islands are claimed by more than one country:

(a) South Shetlands by Britain, Argentina and Chile;
(b) South Orkneys, South Georgia, South Sandwich and the Shag Rocks by Britain and Argentina.

Claims to Antarctic territory have been laid by Argentina, Australia, Chile, France, Britain, New Zealand and Norway. Those of Argentina, Britain and Chile show considerable overlap. Furthermore, Argentina and Chile have pressed their claims further by declaring 200-nml EEZs off their sectors. Countries which have had research bases in the post-World War II period are: Argentina, Australia, Chile, France, Britain, Japan, New Zealand, Poland, South Africa, the Soviet Union (now withdrawing) and the United States.

Environmentally and scientifically, Antarctica is

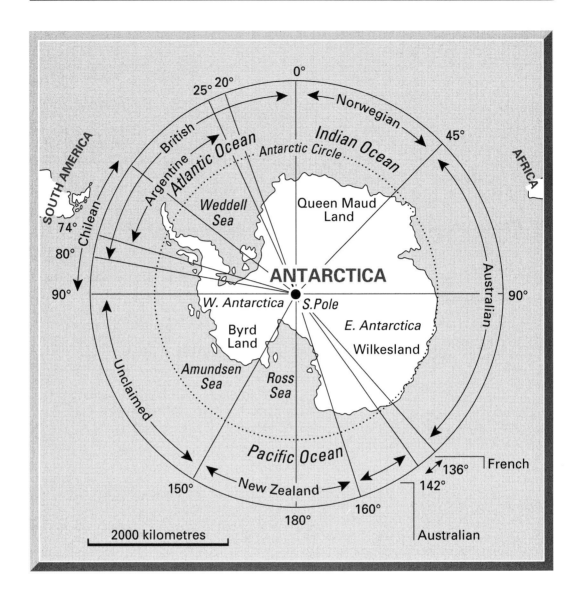

vital as the last great wilderness. Together with the Southern Ocean, it has vast renewable resources, in particular, the highly productive krill, a cheap and plentiful source of protein, which is at present harvested by the Soviet Union, Japan, Chile, Norway, Poland, South Korea, Germany and Taiwan.

Apart from possible problems of over-fishing, the main area of contention is over non-renewable resources. There are manganese nodules on the sea-bed which might yield a range of strategic metals, but, perhaps more importantly, there are an estimated 50 billion barrels of oil, a figure equivalent to 15 per cent of Middle Eastern reserves. With the improved techniques now being developed for sea-floor mining, these resources can no longer be considered unviable. Even more contentious are the strategic minerals thought to lie underneath the ice. Surveys for the United States government have produced very promising results, but the mining process would, of course, be highly destructive of the environment.

Status

By 1991, 32 countries were party to the 1961 Antarctic Treaty. It froze all territorial claims for 30 years, banned all military activity and designated the continent a nuclear-free area. Thus, the Treaty was open for revision in 1991, but delegates to the Madrid conference on Antarctica's future agreed to a 50-year prohibition on mineral mining and prospecting. The agreement states that within this 50-year term, prohibition can only be lifted by a consensus among the 26 voting members of the original Treaty. The ban will run indefinitely unless 75 per cent of those members vote against it and therefore it seems likely that, at least in the medium term, the potential for resource geopolitics has been removed. However, problems remain concerning claims, EEZ policy and possible military activity.

References

Boundary Bulletin, No. 1, (1990), International Boundaries Research Unit, Durham University.
Boundary Bulletin, No. 2, (1991), International Boundaries Research Unit, Durham University.
Boyd, A. (1991), *An Atlas of World Affairs*, Routledge, London.
Child, J. (1985), *Geopolitics and Conflict in South America*, Praeger/Hoover Institution Press, Stanford.
Day, A.J. (ed.) (1984), *Border and Territorial Disputes*, Longman, London.
The Economist Atlas (1989), Economist Books, Hutchinson.
The Economist (1988), 28 May.
The Economist (1989) 20 May.
The Economist (1991) 4 April.
The Economist (1991) 4 May.
The Guardian (1991) 8 August.
Kesteven, G.L. (1978), 'The Southern Ocean', in E.M. Borgese and N. Ginsburg (eds), *Ocean Yearbook, 1*, University of Chicago Press, Chicago.
Morgan, J.R. (1989), 'Naval Operations in the Antarctic Region: A Possibility?' in E.M. Borgese and N. Ginsburg (eds), *Ocean Yearbook, 8* , University of Chicago Press, Chicago, pp. 362–77.
Morris, M.A. (1986), 'E.E.Z. Policy in South America's Southern Cone', in E.M. Borgese and N. Ginsburg (eds), *Ocean Yearbook 6*, University of Chicago Press, Chicago, pp. 417–37.
Morris, M.A. (1988), 'South American Antarctic Policies' in E.M. Borgese and N. Ginsburg (eds), *Ocean Yearbook 7*, University of Chicago Press, Chicago, pp. 356–71.

16 The Aozou Strip

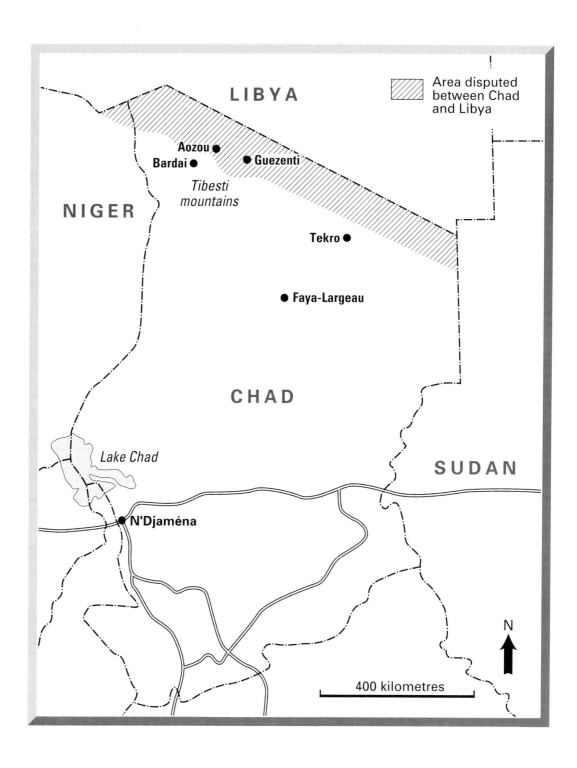

6 The Aozou Strip

Description

The disputed territory comprises a strip along the length of Chad's northern border with Libya. It is approximately 200 km wide and 1,050 km long and is entirely within the present territory of Chad. Within the 14,000 km² of the Aozou Strip, there are said to be valuable resources, particularly of uranium and iron ore. However, the western end is largely occupied by the Tibesti mountains, a volcanic range of strategic significance. North of the range are the sources of the Great Manmade River (GMR) and most of Libya's oil wealth.

History and importance

The Libya–Chad boundary was first agreed under an Anglo–French declaration (13 June 1899), aimed at delimiting the spheres of influence of the two great colonial powers in the region. The agreed alignment was accepted by Italy in an exchange of notes on 1 November 1902, an important development since Italy took Libya from Turkey in 1911. In 1919, an Anglo–French convention determined the point of intersection of the line with the 24°E meridian at 19°13'N.

An agreement of 7 January 1935 between France and Italy was to have ceded the area, now designated the Aozou Strip, to Italy. However, France never ratified the agreement. The Libyan case rests upon the validity of the 1935 agreement, which Chad rejects on the grounds that it was never ratified or brought into force. In fact, Chad cites the Treaty of Friendship and Good Neighbourliness between France and Libya (then independent) of 10 August 1955 as the continuing basis for the boundary in its present position.

Taking advantage of factional fighting in Chad, Libya occupied the Strip in 1972 and in June 1973 officially annexed it. It was hoped that uranium deposits, similar to those which occur in neighbouring Niger, would be discovered and the annexation was widely interpreted as indicating such a discovery. As a result, it was thought that ownership of the Strip would allow Libya to have nuclear independence, becoming eventually a nuclear power. Relations between Libya and Chad later improved and it was even intimated that the Strip had been sold back to Chad for $120 million, a charge denied in Chad. In 1976, Chad denounced the Libyan occupation as well as Libyan support for anti-government rebels. The Organization of African Unity (OAU) subsequently set up a commission on the dispute (July 1977).

Following the resignation of the president, a government of national unity was established in Chad in 1979, and on 15 June 1980 a Treaty of Friendship was signed with Libya. President Oueddi of Chad accepted the annexation of the Aozou Strip as an 'accomplished fact' on 26 December 1980. However, the move towards a full unity between the two countries proposed by Libya's Muammar Quddafi was opposed by Chad, and Libya again became embroiled in a series of Chadian civil wars throughout the 1980s. In 1987, Chadian forces succeeded in defeating the Libyans and driving them back into the area around Aozou itself, before the OAU mediated a cease-fire. Libya's prestige was badly damaged, but by 1989 the Chad regime itself began to disintegrate. A rebel group led by Idriss Deby set up its headquarters in neighbouring Sudan and in November 1990, re-entered the country to overthrow the president and his government.

Status

After five rounds of talks between senior Libyan and Chadian ministers had failed to settle the dispute, the two countries agreed to submit the case to the International Court of Justice (the World Court). Libya instituted proceedings against Chad on 31 August 1990 and Chad against Libya on 1 September 1990. The date for the presentation of memorials (volumes of evidence) was set for 26 August 1991 and for counter-memorials on 27 March 1992. The case appears to rest on the interpretation of the 1935 Treaty, the presence or otherwise of minerals and the strategic importance of the region. If Libya is to have a defensible boundary, some re-alignment may be required and it may be possible that there could be a satisfactory exchange of territory.

References

Blake, G.H. and Drysdale, A. (1985) *The Middle East and North Africa: A Political Geography*, Oxford University Press, Oxford.

Boundary Bulletin, No. 1, (1991) International Boundaries Research Unit, Durham University.

Brownlie, I. (1979), *African Boundaries: A Legal and Diplomatic Encyclopaedia*, Royal Institute of International Affairs, London.

Day, A.J. (ed.) (1984), *Border and Territorial Disputes*, Longman, London.

Downing, D. (1980), *An Atlas of Territorial and Border Disputes*, New English Library, London.

International Court of Justice Communiqué (1990), (unofficial – for immediate release), 4 September.

International Court of Justice Communiqué (1990), (unofficial – for immediate release), 2 November.

International Court of Justice Communiqué (1991), (unofficial – for immediate release), 27 August.

United States Department of State (1978), *Chad – Libya*, International Boundary Study No. 3 (revised), Office of the Geographer, Bureau of Intelligence and Research, Washington DC, 15 December.

7 The Attila Line

Description

Originally a cease-fire line, following the Turkish invasion of northern Cyprus in 1974, the Attila Line is now a *de facto* boundary. It extends for 180 km across the island from the Kokkina enclave (west of Xeros) in the west, through the capital of Nicosia (Lefkosa) to immediately south of Famagusta (Gazi Magosa) in the east. Between the Greek and Turkish cease-fire lines, the UN Peace-keeping Force in Cyprus (UNFICYP) controls a buffer zone varying in width from 7 km at its greatest extent to 20 m at its narrowest point within Nicosia (Lefkosa). This buffer zone accounts for approximately 3 per cent of the total area of Cyprus.

History and importance

Predominantly Greek since antiquity, from 1571 to 1878 Cyprus was under Turkish rule and it was during this period that the minority Turkish population was introduced. The island was ceded to Britain in 1878, and from the 1930s, elements of the Greek majority began demanding 'Enosis' or union with Greece. From 1955 to 1960, the National Organization of Freedom Fighters (EOKA), fought a bitter guerrilla war against the British. This resulted in the London–Zurich tripartite agreements between Britain, Greece and Turkey of 1959 and independence in August 1960. The Turks were offered constitutional guarantees, including the permanent vice presidency and vetoes over certain legislation. By the Treaty of Establishment, Britain retained two soverign bases, Dhekelia and Akrotiri, covering some 256 km^2.

Both Greece and Turkey were allowed to station military contingents on the island. At the time of the last reliable census (1961), the Turks were outnumbered by the Greeks in the ratio 4:1, and the Greek population amounted to approximately 650,000. At that stage, the islands had some 110 mixed villages, together with distinct population enclaves in most towns.

For the first three years after independence, there was peaceful cooperation, then on 30 November 1963, the Greek-Cypriot president (Makarios) proposed 13 amendments to the constitution, effectively demoting the Turkish Cypriots from co-founder status and paving the way towards 'Enosis'. As a result, on 28 December 1963, the Green Line was established in Nicosia to separate the warring factions. However, inter-communal violence spread and approximately 25,000 Turkish Cypriots moved into enclaves, defended by their own paramilitary forces.

On 27 March 1964, UNFICYP took over from British troops, but tensions mounted, reinforced from 1967 by the rise of a military junta in Greece. In 1974, the junta masterminded a *coup d'état* against President Makarios but failed to kill him, and Nicos Samson, a former EOKA gunman, was installed as head of a regime of 'national salvation'. As a result of these events, on 20 July 1974, five days after the assassination attempt, Turkey exercised its right to intervene under Article 4 of the Treaty of Guarantee and launched a military invasion. Subsequent peace talks collapsed and there was a second Turkish advance from 13 August, resulting eventually in the cease-fire line agreed on 18 August 1974.

Following the establishment of the Attila Line, the Turks occupied 37 per cent of the island, and between July 1974 and December 1975, a massive population exchange, comprising 185,000 Greek Cypriots and 145,000 Turkish Cypriots, took place. Northern Cyprus was designated the 'Turkish Federated State of Cyprus' on 13 February 1975, but received no international recognition except from Turkey.

Status

Without special permission and supervision, civilian movement across the Attila Line is restricted, but,

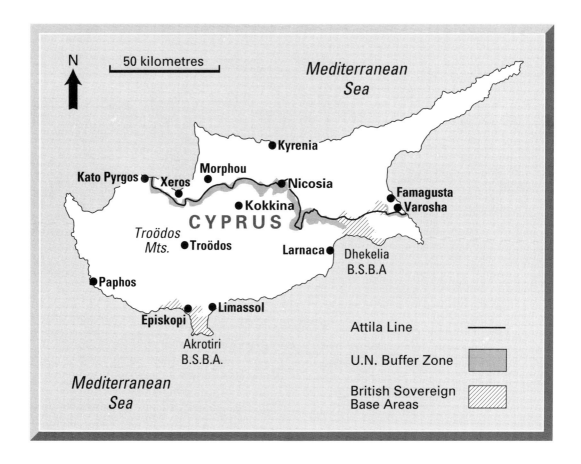

in the buffer zone itself, under the UN-designated Farming Security Line, farmers from both sides are allowed to use the land. However, the Line divides important agricultural areas, disrupts communications and bisects water resources. Furthermore, it represents the focus of confrontation between Greece and Turkey and, in combination with tensions in the Aegean, Cyprus could become a regional flashpoint for conflict.

References

Blake, G.H., Dewdney, J., and Mitchell, J. (1987), *The Cambridge Atlas of the Middle East and North Africa*, Cambridge University Press, Cambridge.

Boyd, A. (1991), *An Atlas of World Affairs*, Routledge, London.

Day, A.J. (ed.) (1984), *Border and Territorial Disputes*, Longman, London.

The Economist (1991) 27 July.

The Economist (1991) 10 August.

Hitchens, C. (1990), 'The Island Stranded in Time' in *Frontiers*, BBC Books, London, pp. 116–41.

Mirdagheri, F. (1991), 'Hopeful signs but deadlock continues', in *Boundary Bulletin*, No. 1, International Boundaries Research Unit, Durham University, pp. 13–14.

Munro, D. and Day, A.J. (1990), *A World Record of Major Conflict Areas*, Arnold, London.

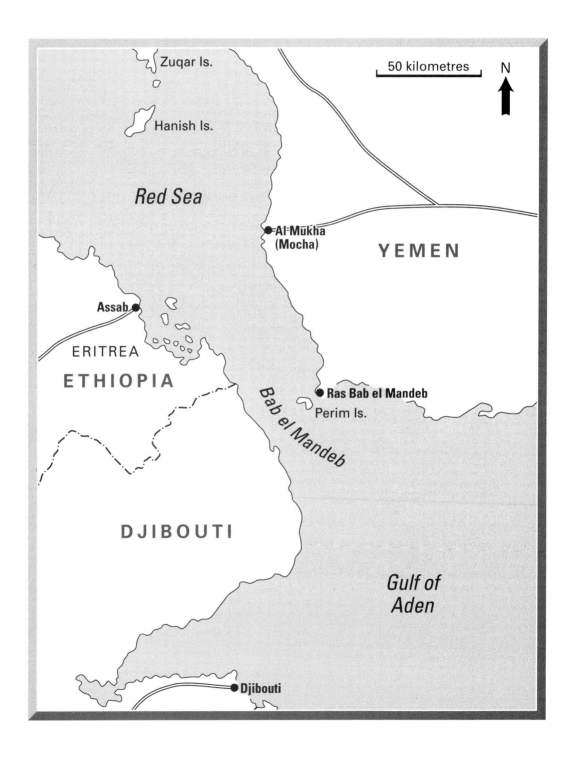

8 Bab el Mandeb

Description

Located at the southern entrance to the Red Sea, Bab el Mandeb ('Bab' is Arabic for Strait) is one of the major choke points of the world. A choke point is a location where, for physical reasons such as the constriction of a sea-line of communication between land masses or as a result of economic necessity, shipping becomes concentrated. This concentration is clearly vulnerable to attack from the land, the air, the sea or under the sea and choke points are therefore seen as nodes of vulnerability along sea-lines. Even if military action is unlikely, threats of such action can cause problems.

Bab el Mandeb itself is divided by Perim Island (8 km^2) into a small strait to the east (the Yemeni side) and a larger strait to the west (the African side). The main shipping route to the west of Perim Island is approximately 16 km wide and 100–200 m deep. The strait is in a particularly volatile location, with the recently reunited Yemen on one side and Ethiopia and Djibouti on the other.

History and importance

Perim Island was occupied by Britain in 1799, while the nearby port of Aden on the Arabian peninsula was annexed by Britain in 1839 as part of a string of naval bases established to protect the empire in India. Britain imposed a wholly artificial boundary between north and south Yemen to facilitate its control over Aden and its hinterland. The southern part was known as the Aden Protectorate and this was retained after Indian independence in 1947. Even following the war of 1956 and the loss of the Suez Canal, Britain retained Aden as a base for operations in the Gulf. However, given Arab opposition, the position eventually became untenable and Britain withdrew in November 1967.

Two incidents are usually cited as illustrating the strategic significance of Bab el Mandeb. On 11 June 1971, the *Coral Sea*, a Liberian-flag tanker chartered by Israel, was attacked in the Strait of Tiran by rockets from a gunboat, positioned in the vicinity of Perim Island. Later, in July and August 1984, some 18 vessels were damaged by mines placed at the northern and southern ends of the Red Sea. Responsibility was never established, but suspicion focused on Libya.

The importance of Bab el Mandeb is related to traffic through the Suez Canal, the most important of which by far being tanker traffic between the Gulf and Europe, but this has been undermined by the Saudi Petroline, now capable of taking 5 million bpd. However, this may be partly counteracted by the widening of the Suez Canal and the further development of the SUMED pipeline. Although no official statistics have been taken, it is thought that some 20,000 ships annually, or about 55 per day, transit the strait, thus making it among the more heavily used sea lanes in the world. During the closure of the Suez Canal, from 1967–75, the Cape Route round South Africa came into prominence, but it has since declined considerably in importance. Statistics compiled by the South African navy indicate that fewer than 6,000 ships a year now sail round the Cape.

Bab el Mandeb is also vital to a number of Red Sea states, particularly Sudan, Jordan and Israel. It has also, in the past, been the scene of superpower rivalry, with, at various times, Soviet facilities being established at Aden and Berbera, Somalia, and on Socotra Island 650 miles to the east. However, from 1978, the United States replaced the Soviet Union in Berbera. Djibouti, an independent state with a very strong French influence, remains the headquarters of the French Indian Ocean navy and, during Operation Desert Storm was a staging post for some 3,800 troops.

Status

Bab el Mandeb is vital to the world trade infrastructure and it is of unquestionable strategic importance. Nevertheless, with the decline of East–West confrontation, there has come a series of major changes in the region. On the Arabian side, the reunification of Yemen has eliminated conflict, but on the African side there has been continuing unrest in Ethiopia (Eritrea) and northern Somalia, while even Djibouti now appears less stable.

References

Anderson, E.W. (1985), 'Dire Straits', *Defense and Diplomacy*, 3, No. 9, pp. 16–20.

Blake, G.H., Dewdney, J. and Mitchel, J (1987), *The Cambridge Atlas of the Middle East and North Africa*, Cambridge University Press, Cambridge.

Blake, G.H. and Drysdale, A. (1985), *The Middle East and North Africa: A Political Geography*, Oxford University Press, Oxford.

Cottrell, A.J., and Hahn, W.F. (1978), *Naval Race or Arms Control in the Indian Ocean*, Agenda Paper No. 8, National Strategy Information Center, Inc., New York.

Lapidoth, R. (1975), *Freedom of Navigation with Special Reference to International Waterways in the Middle East*, The Hebrew University, Jerusalem.

Peterson, J.E. (1985), 'The Islands of Arabia: Their Recent History and Strategic Importance' in Arabian Studies, **VII**, London.

Rais, R.B. (1986), *The Indian Ocean and the Superpowers*, Croom Helm, London.

9 The Baltic Republics

Description

The 'Baltic Republics', as they are popularly known, comprise a group of small states, each different from the others, united through common political, economic and social concern. Despite the small size of their populations, each speaks a distinctive language and has its own heritage. The basic statistics are as follows:

	Area km^2	Population
Estonia	45,100	2 million
Latvia	63,700	3 million
Lithuania	65,200	4 million

A key difference, and one that is likely to cause increasing problems as a result of conflict within the former Soviet Union, is the ethnic mix of each. Many Russians were settled in the three Baltic Republics, in many cases replacing deported inhabitants, and by the 1980s Russian speakers made up 40 per cent of Estonia's population, almost 50 per cent of Latvia's, but only 15 per cent of Lithuania's. Estonia and Latvia were incorporated into the Swedish Empire in the 17th century, but as a result of the Great Northern War (1700–21) were transferred to Russia as was Lithuania, subsequent to partitionings of Poland in the 18th century.

History and importance

Following the 19th century rise of nationalism and independence, and a weak Russia and Germany, a re-emergent Poland seized the Lithuanian capital, Vilnius, in 1918 and annexed Memel, an old Hanseatic port, in 1923. Memel was subsequently returned in 1939 after an ultimatum from the Nazi German government, while Vilnius, together with approximately one-sixth of the total area of the present Lithuania was reclaimed by the Soviets in their take-over of 1939. Having been independent between the two World Wars, the three Baltic States were occupied by the Soviet Union and incorporated into the USSR in 1940. This was confirmed by plebiscites in 1944, although each state claimed that its plebiscite was a fraud when declaring independence in 1990.

Thus, the Baltic States have witnessed, throughout their histories, conflict and competition among their more powerful neighbours. Occupying a key situation in the Baltic, their strategic significance is still apparent. Following their independence, the Soviet Union lost not only the important land bases, but the vital ice-free ports of Riga and Tallin.

Status

Lithuania declared independence on 11 March 1990 and Estonia and Latvia followed in May 1990. All have received international recognition. Indeed, several states, including the United States and Britain, never recognized the Soviet accession.

Apart from the minorities, comprising mainly militant Russians, the major problems of the Baltic States are economic. This has been acknowledged, particularly by the Scandinavian countries, with whom closer ties have been rapidly forged. Indeed, in November 1990 the 'Baltic Family' Conference was held to confirm and strengthen such ties. None the less, all three remained heavily reliant upon the former Soviet Union and have signed new agreements with the Commonwealth of Independent States (CIS). The other important geopolitical aspect concerns the future of Kaliningrad, an isolated part of Russia. The city itself hopes to become a free-trade zone in which the largest port in Russia would be built. This would be designed to handle at least 25 million tonnes per year and would act as the 'Hong Kong' of the Baltic.

References

Anderson, E.W. (1991), 'Lithuania 1990' in *Boundary Bulletin*, No. 1, International Boundaries Research Unit, Durham University, p. 13.

Berdichevsky, N. (1991), 'A Tale of Two Cities: Lithuania's See-Saw Struggle to Gain Control of Vilnius and Klaipeda (Vilna and Memel)' in *Boundary Belletin*, No. 2, International Boundaries Research Unit, Durham University, pp. 5-9.

Boundary Bulletin, No. 2 (1991), International Boundaries Research Unit, Durham University.

The Economist (1991) 12 January.

The Economist (1991) 27 April.

The Economist (1991) 3 August.

The Economist (1991) 31 August.

The Economist (1991) 7 September.

The Independent, October, 1991.

Munro, D. and Day, A.J. (1990), *A World Record of Major Conflict Areas*, Arnold, London.

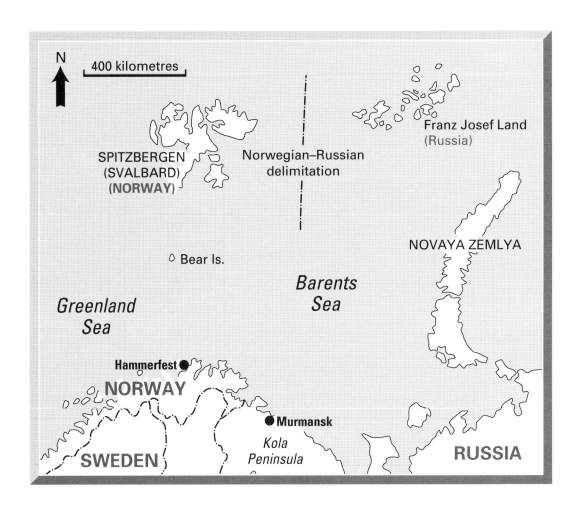

10 The Barents Sea

Description

The Barents Sea is a marginal sea of the Arctic Ocean. It is bounded by the island of Novaya Zemlya to the east, the Greenland Sea to the west, the White Sea, Russia, Norway and the Norwegian Sea to the south and Spitzbergen (Svalbard) and the archipelago of Franz Josef Land to the north. It is relatively shallow except for a narrow trench, 100 m or more deep, close to the mainland. The effects of the North Atlantic Drift, a warm current, are still felt this far north, producing salinity and temperature contrasts and an ice-free zone as far east as the Russian port of Murmansk.

History and importance

This sea has long been considered vital to Soviet (and now Russian) interests, since its main offensive force, the Northern Fleet, is based primarily at Polyarny and Murmansk and is controlled from Severomorsk, all three places being on the Tuloma River. The prevalence of deep water inshore allowed, in particular, the ballistic-missile nuclear submarines to slip round the Northern Cape and into the Greenland Sea. To protect them, there are about 40 airfields in all in the Kola Peninsula. With the end of the Cold War, the threat may have waned, but the Kola Peninsula remains the most heavily armed area in the world.

If the importance of defence issues has declined, that of resources remains. The Barents Sea yielded approximately 20 per cent of the annual Soviet fish catch, and there is great potential for exploitable oil and natural gas in the sea-bed. Norway and the Soviet Union (whose claims have now been assumed by Russia) dispute the rights of exploitation. Indeed, this was the only active territorial dispute between countries of NATO and the Warsaw Pact during the Cold War. The Soviet Union contended that the line of demarcation should be drawn due north from the Soviet–Norwegian land border, while the Norwegian claim is based on a median line, constructed between Spitzbergen and Novaya Zemlya. The result is that there are two disputed areas, comprising some 180,000 km^2, and for the purposes of fishing and security, both countries declared the contested waters a common 'Grey Zone' (1978).

The dispute over Spitzbergen is considered elsewhere (Map 64). That over Franz Josef Land, 600 nml north of North Cape and 200 nml from Novaya Zemlya, was settled long since; the Soviet Union proclaimed ownership in 1926 and this was disputed by Norway as late as 1930, but since then the status has remained unchallenged.

Status

Given the military importance of the region, and the volatile state of the former Soviet Union, it is not possible yet to consider the Barents Sea as strategically unimportant. It still represents the route by which, either by submergence in deep water or underneath the northern ice, the most powerful ships ever built can move into the world's oceans. Furthermore, as offshore mineral exploitation becomes a reality, the need for an accepted demarcation between Norway and Russia will become a necessity. However, the preoccupation of the Russian Republic with domestic, economic, ethnic and political problems must surely mean that an active prosecution of its territorial claims will be shelved for at least the time being. The Barents Sea might best therefore be termed, using the parlance of vulcanicity, a 'dormant' flashpoint but which, at some point in the future, will certainly erupt again.

References

Boyd, A. (1991), *An Atlas of World Affairs*, Routledge, London.

Leighton, M.K. (1979), *The Soviet Threat to N.A.T.O's Northern Flank*, Agenda Paper No. 10, National Strategy Information Center, Inc., New York.

Luton, G. (1986), 'Strategic Issues in the Arctic Region' in E.M. Borgese and N. Ginsburg (eds), *Ocean Yearbook 6*, University of Chicago Press, Chicago, pp. 399–416.

Prescott, J.R.V. (1985), *The Maritime Political Boundaries of the World*, Methuen, London.

11 The Basque Country/Euzkadi

Description

The Basque country is the coastal area of the eastern Pyrenees, which straddle both sides of the Franco-Spanish border. The Basques themselves are a people of obscure origins who have preserved a cultural identity based on language and history. Mainly as a result of economic development, the Spanish Basque country, in marked contrast to its French counterpart, has become a place of conflict. The Spanish Basque country comprises four provinces: Alava, Guipuzcoa, Navarra and Vizcaya. These four provinces have a total area of 17,700 km^2 and a population of approximately 2.5 million. Since 1980, part of the Basque country has been recognized as an autonomous region. This portion, comprising Alava, Guipuzcoa and Vizcaya, has an area of 7260 km^2 and a population of about two million.

History and importance

Basque nationalism arose during the latter part of the 19th century as a result of political centralization and rapid industrialization. Although Spain was officially unified within its present boundaries in 1715, there were strong centrifugal forces and many peripheral areas remained powerful. For example, the Basques, retained their own legal-administrative system, which included the communal ownership of some natural resources, a separate taxation system and their own tariff frontiers. As the powers of the central government increased and moves towards standardization resulted in the dismantling of such local structures, regional tensions mounted. In the case of the Basque country, by the late 19th century, it had lost its main 'autonomous' features and the language was in sharp decline.

At the same time, industrialization occurred at a spectacular pace, resulting in a large influx of Spanish-speaking workers. By the early years of the 20th century, Bilbao was responsible for approximately 20 per cent of the total world production of iron ore. Thus, the Basque region became distinctive in what was still predominantly an undeveloped agricultural state. Basque nationalism was therefore a reaction against a number of fundamental changes which coincided: the influx of Spanish workers, the emergence of socialism, appalling industrial pollution and the loss of traditional Basque practices.

The first Basque nationalist newspaper appeared in 1893, and in 1895 the nationalists of Bilbao founded the PNV (Partido Nacionalista Vasco), which had as its main slogan, 'God and the Old Laws'. Nationalist aspirations were briefly enhanced during the Second Republic, only to be brutally dashed during the Civil War. On behalf of General Franco, Guernica, the ancient capital, with its famous oak tree, the cultural symbol of Basque autonomy, was bombed by the Luftwaffe and over 15,000 people were killed. The incident was immortalized in a painting by Picasso which remained a major symbol of resistance throughout the Franco regime and was only returned to Spain on the resumption of the monarchy after Franco's death.

Repression of Basque culture followed the Civil War, with the language and folk music outlawed. The militant and separatist ETA (*Euzkadi ta Askatasuna*, Basque Homeland and Freedom) was formed in 1959 as a breakaway group from the moribund PNV. During the Franco period, it developed a radical brand of Basque nationalism and, as a result of guerrilla operations, transformed the political situation. Its military campaign provoked severe repression and in 1970, the trials of 16 ETA activists at Burgos sparked off protest strikes and directed global attention to Spain. ETA's most significant action was its assassination of General Franco's chosen successor, Admiral Carrero Blanco in 1973, which upset plans for the continuation of authoritarian rule and enhanced the pace of democratic transition. With the death

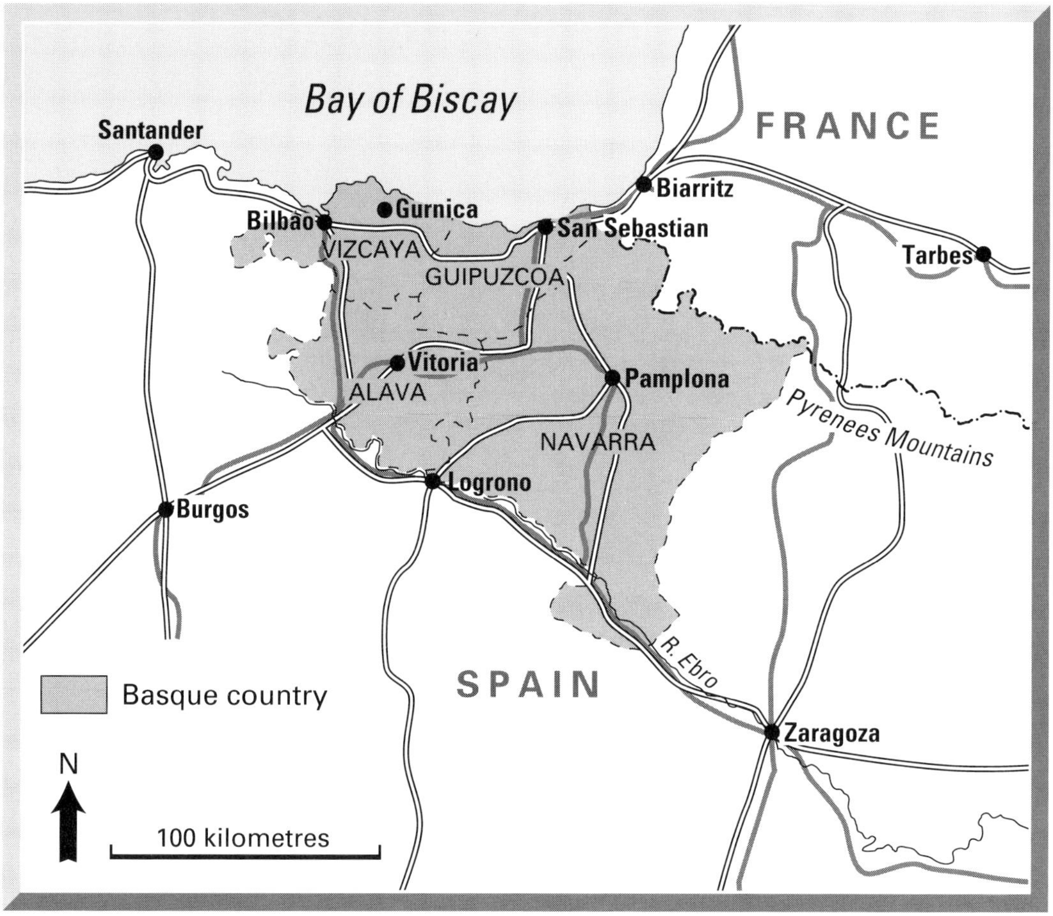
Bay of Biscay
FRANCE
Santander
Biarritz
Gurnica
Bilbao
San Sebastian
Tarbes
VIZCAYA
GUIPUZCOA
Vitoria
Pamplona
ALAVA
Pyrenees Mountains
NAVARRA
Logrono
Burgos
R. Ebro
Basque country
SPAIN
Zaragoza
N
100 kilometres

of General Franco on 20 November 1975 and the subsequent return of Spain to democracy, ETA lost a large part of its impetus. It became further marginalized after the elections of 1980 when the more moderate nationalist groups came to dominate the Basque Assembly. The result was a limited autonomy for three provinces.

The importance of the continuing, if now relatively minor, conflict in the Basque country is that it remains the only serious flashpoint in Spain. Many peripheral areas in Western Europe exhibit centrifugal tendencies, but few are afflicted by the level of violence which has, at times, erupted in the Basque country.

Status

The objective of ETA remains an independent state, Euzkadi, for the Basque nation. This would comprise not only the four Spanish provinces, but also the three adjacent Basque provinces of Bas-Navarre, Lapurdi and Civeroa in France. The slogan remains 'Seven in One', but support for the movement has declined to approximately 15 per cent of the population of the Spanish Basque country. However, confrontation between the centre and the periphery continues and ETA, now linked to a separatist coalition, Herri Batasuna (United People) is still a force. With the integration of Europe, there may well be a future for small autonomous areas as separate units, but the creation of the Single European Market in the European Community at the end of 1992 will also result in increasing centralization. The Basque country is likely therefore to remain only a local flashpoint but one which, given the increasing global importance of Europe, will occasionally enjoy a high profile.

References

Calvocoressi, P. (1991), *World Politics Since 1945*, sixth edition, Longman, London.

Chisholm, M. and Smith, D.M. (eds) (1990), *Shared Space: Divided Space*, Unwin Hyman, London.

Johnston, R.J. and Taylor, P.J. (eds) (1986), *A World in Crisis?* Blackwell, Oxford.

12 The Beagle Channel

Description

The Beagle Channel is the southerly of two channels which link the Atlantic and Pacific Oceans in the far south of South America. The only other ways of making the inter-oceanic passage are via the Panama Canal or Cape Horn. The Beagle Channel is narrow and sinuous but sheltered. Approximately half of its course is accepted as the international boundary between Argentina and Chile, while the western half is totally within Chilean territory. The channel is named after *HMS Beagle*, the British survey ship in which Charles Darwin made his famous voyage in the 1830s. Although the dispute between Argentina and Chile is known as the 'Beagle Channel dispute'. it is not the channel itself which is contested, but three small islands at the eastern end of the channel and south of Tierra del Fuego: Lennox, Nueva and Picton.

History and importance

The international boundaries in the area were originally recognized in 1810, but, following disputes in the 1870s, there was a new Boundary Treaty of 23 July 1881. This assigned all islands east of Tierra del Fuego to Argentina and all islands south of the Beagle Channel, as far as Cape Horn and west of Tierra del Fuego, to Chile. There were, however, differences over demarcation and the British government, which had been in charge of executing the 1881 treaty, was asked, in 1896, to arbitrate.

In fact, the 1890s saw an arms race in both Chile and Argentina and there was a distinct possibility of war over the issue in 1898. Presidents of both countries met in 1899 and in 1902 both governments subscribed to a General Treaty of Arbitration for the solution, by friendly means, of any bilateral problems. The British Crown was made responsible for arbitration. The dispute resurfaced in 1904, 1905, 1915, 1933, 1938, 1955, 1964 and, finally in 1967, when on 7 December, Chile lodged an appeal based on the General Treaty of Arbitration (1902). Britain took up its responsibility and asked four other members of the International Court of Justice (ICJ) (France, Nigeria, Sweden and the United States) to set up an Arbitration Court.

In February 1977, Britain formally accepted the Arbitration Court's decision, which was approved by the Queen in April and delivered to Chile and Argentina on 2 May. According to the decision reached, Chile would retain the three islands and other islands further south, including that on which Cape Horn is situated. On the same day, Chile accepted the terms, but later, on 25 January 1978, Argentina rejected them, issuing a 'nullity declaration'. During the latter part of 1978, war over the issue seemed a strong possibility as both countries mobilized and built up their forces in the disputed south. Conflict was only forestalled when it was agreed to refer the dispute to the Pope.

Status

The 'Beagle Channel dispute' is interesting, not only for its longevity, but because the relative importance of the area has changed. Originally, the dispute was over territory and, possibly, bases, but with the introduction of the EEZ, the problem was reinforced. Possession of the islands could determine the control of large areas of ocean with their attendant resources of fish and sea-floor minerals. The proposals presented by Pope John Paul II in 1980 resembled those of the ICJ judges and were, again, refused by Argentina. However, later Papal pressure, combined with Argentina's military defeat in the Falklands and Chile's desire to avoid a full-scale war, led to compromise and the 1984–5 Peace and Friendship Treaty. The Pope's original terms were modified slightly, but in essence, the decision was the same. Chile was to keep the islands and a boundary running due south from Cape Horn was accepted. Although, effectively, Argentina lost the case, there were sufficient adjustments with regard to navigation and access that Argentina's geopolitical and maritime aspirations may well have been satisfied. Clearly, the question of the Malvinas looms far larger.

References

Boyd, A. (1991), *An Atlas of World Affairs*, Routledge, London.

Child, J. (1985), *Geopolitics and Conflict in South America*, Praeger/Hoover Institution Press, Stanford.

Day, A.J. (ed.) (1984), *Border and Territorial Disputes*, Longman, London.

Downing, D. (1980), *An Atlas of Territorial and Border Disputes*, New English Library, London.

Morris, M.A. (1986) 'E.E.Z. Policy in South America's Southern Cone in E.M. Borgese and N. Ginsburg (eds), *Ocean Yearbook 6*, University of Chicago Press, Chicago, pp. 417–37.

Morris, M.A. (1988) 'South American Antarctic Policies' in E.M. Borgese and N. Ginsburg (eds), *Ocean Yearbook 7*, University of Chicago Press, Chicago, pp. 356–71.

Prescott, J.R.V. (1985), *The Maritime Political Boundaries of the World*, Methuen, London.

Santis-Arenas, H. (1989), 'The Nature of Maritime Boundary Conflict Resolution between Argentina and Chile, 1984 in *International Boundaries and Boundary Conflict Resolution*, 1989 Conference Proceedings, International Boundaries Research Unit, University of Durham, pp. 301–22.

13 The Benguela Railway

Description

The need for a railway to transport minerals out of Central Africa was originally perceived in 1902 by a British entrepreneur, Robert Williams. In 1931, the Benguela Railway opened, linking the ports of Lobito and Benguela, on the Angolan coast, with central Angola, Shaba Province in Zaïre and Ndola in central Zambia. This stretch of some 2,500 km forms roughly the western half of the trans-African railway from Lobito to Beira, Mozambique, in the south, and Dar-es-Salaam in the north (see Tanzam Railway). Before the line was severed by National Union for the Independence of Angola (UNITA) guerrillas in 1975, it was carrying over half the copper exports of Zambia.

History and importance

Following the revolution of 1974 in Portugal, Angola, the main Portuguese territory in Africa, achieved independence on 11 November 1975. However, three rival guerrilla movements were left fighting in a civil war that lasted until 1991. The Marxist–Leninist, Soviet- and Cuban-backed Popular Movement for the Liberation of Angola (MPLA) gained control of the central government in Luanda. In the north, based on the Bakongo tribes, was the National Front for the Liberation of Angola (FNLA) and in the south, based on the Ovimbundu tribe, was UNITA. During the 1980s, FNLA gradually disintegrated, but UNITA was able to conduct an effective and long-running campaign, directed by its charismatic leader, Jonas Savimbi, against the MPLA. Early success in closing the Benguela line was followed by many setbacks, largely as a result of some 50,000 Cuban troops, flown in by the Soviet Union.

There was an attempt to bring peace as a result of UN Resolution 435 (1978), but it was only following the Gorbachev reassessment of Soviet strategy that the resolution could be implemented in December 1988. After United States' mediation, Angola, Cuba and South Africa signed the agreement under which there was to be a South African military withdrawal from Namibia in 1989 and a phased departure of Cuban troops from Angola by mid-1991. A temporary UN peace-keeping force took over in Namibia, which became independent, with a (SWAPO) South West Africa People's Organization-based government in 1990. Thus, although the fighting continued in Angola in 1990, the government (MPLA) had lost its Cuban support and UNITA its backing from Namibia-based South African troops, leading to a cease-fire in May 1991. The reconciliation between UNITA and MPLA began with a cease-fire and the Gbadolite Declaration of June 1989.

Thus, for virtually a quarter of its existence, the operation of the Benguela Railway was bedevilled by the Angolan civil war. Therefore, its great potential economic and strategic importance to the central southern African states and particularly Zambia has been largely unrealized. The copper and the strategically more important cobalt, produced in Shaba Province and northern Zambia, could have been exported from Angola, thereby avoiding transit through South Africa and transport on the most vulnerable section of the Cape Route.

In the event, the minerals were exported in containers via South African railways from East London, enabling South Africa, with at least three other key strategic metals – chromium, manganese and platinum – to maintain a strangle-hold on a fourth, cobalt. This considerably reduced the power of the Front Line States in opposing apartheid and heightened the perceived strategic importance of South Africa in Western thinking.

Status

It has been reported recently that the western half of the Benguela Railway is now open to traffic and

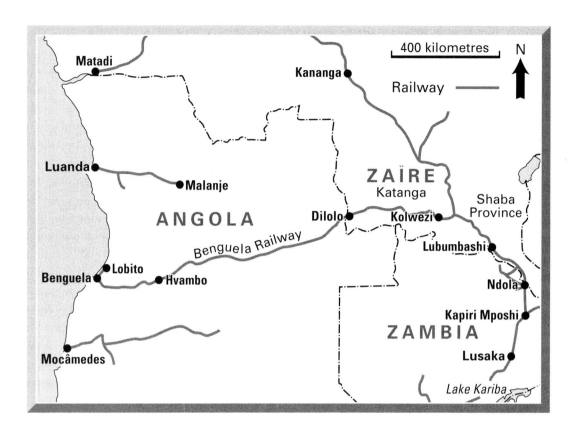

restoration is being continued on the remainder of the line. When it is again possible to export cobalt from Zaïre and Zambia, the geopolitical status of South Africa will have been further diminished and the land-locked areas of central southern Africa will be able to act more independently of their powerful southern neighbour.

References

Anderson, E.W. (1988), *Strategic Minerals: The Geopolitical Problem for the United States*, Praeger, New York.

Boyd, A. (1991), *An Atlas of World Affairs*, Routledge, London.

Griffiths, I.L.-L. (1985), *An Atlas of African Affairs*, Methuen, London.

Munro, D. and Day, A.J. (1990), *A World Record of Major Conflict Areas*, Arnold, London.

14 Berlin

Description

Covering an area of 884 km² and with a population of 3.5 million, Berlin, the pre-World War II capital of Germany, was confirmed in 1991 as the capital of the united Germany. During the post-War period, Berlin was effectively a microcosm of the Cold War world. In location, it is as eccentrically located to the East as Bonn, the former capital, was to the West in the Federal Republic of Germany. Thus, its hinterland to the east is limited by the Oder-Neisse border, a frontier confirmed by Germany in 1990.

History and importance

Towards the end of World War II, Greater Berlin was divided as a result of two Protocols between the Allies, signed in London on 12 September and 14 November 1944, together with the Potsdam Conference (17 July–2 August 1945). The main division was between East Berlin (the Russian Sector) and West Berlin (the Western Sector), the latter being divided into French, British and American sectors. Since the whole city was in East Germany (the German Democratic Republic), West Berlin was effectively an enclave of West Germany (the Federal Republic of Germany). It was linked to the West by air and by an autobahn, but these communications were under constant Soviet threat.

On 23 and 24 June 1948, the major blockade of West Berlin began and on 28 of that month, the United States announced the Berlin Airlift. This continued for almost a year until the blockade was lifted on 12 May 1949. In June 1953, an anti-Communist uprising in East Berlin was crushed with the help of Soviet troops and on 13 August 1961, the border between East and West Berlin was closed.

Four days later, the construction of the Berlin Wall began. This included a strip of no man's land 100 m wide and six crossing points. From 1963, there was a certain easing of tension and travel concessions were introduced, but it was not until 3 September 1971 that, as a result of the Quadripartite Agreement on Berlin, it was confirmed that all disputes must be settled by peaceful means, that the situation could not be changed unilaterally and that transit traffic was to be unimpeded. The map shows the sectorization of the city and the course of the Berlin Wall during the period 1961–89.

On 9 November 1989, the Wall and the Iron Curtain between the two Germanies was opened and in the next three days, over three million East German citizens visited West Germany and West Berlin. By 1 July 1990, when the foundations for an economic and social union had been laid, all border controls were abolished. On 3 October 1990, East Germany (the German Democratic Republic) ceased to exist. It was redivided into the five *Lander* which had been abolished under Communist rule and acceded to the Basic Law of the Federal Republic. These five *Lander* took their place alongside the 10 Western *Lander* and Berlin. In June 1991, the parliament of Germany voted to move the Federal Government to the united Berlin.

Berlin has obviously enjoyed great historical importance, and as the capital of the economically most powerful state in Western Europe this is likely to be reinforced. Indeed, with the return of the countries of Eastern Europe and even the states of the former Soviet Union into the mainstream of European life, Berlin is ideally placed to become the centre of gravity for the new Europe.

Status

While the future looks extremely bright as a world metropolis, linking East and West, the present is fraught with problems. Living standards were far lower in East Berlin and the city is short of at least 150,000 dwellings. West Berlin was modernized and planned as a separate entity and the inclusion of East Berlin necessitates replanning. There are also major problems concerning conservation, particularly in the context of the industrial areas in the East.

Politically, there could also be problems. The December 1990 Berlin Senate Election provided an overall vote of 40.3 per cent for the Christian Democrats, with 48.9 per cent in the West, but only 25 per cent in the East. Indeed, in East Berlin the former Communists received 24 per cent of the vote. Economically, difficulties also loom. In East Berlin, unemployment is three times that in the West and those employed earn, generally, only a half of the average West Berlin income. Thus, before it can develop internationally, Berlin must overcome a host of local problems and in doing so, it will be a sizeable drain on the German economy.

References

The Economist (1991) 9 November.
Day, A.J. (ed.) (1984), *Border and Territorial Disputes*, Longman, London.
Reid, W. (1991), 'A Retrospective Look at the Border Between the Two Germanies', in *Boundary Bulletin*, No. 1, International Boundaries Research Unit, Durham University, pp. 14–16.

15 Bessarabia

Description

Until 1812, Moldova comprised the whole area from the Dniester River westwards, across the River Prut (the current boundary) to the Carpathians. In that year, the Russians annexed Bessarabia, the area between the Prut and the Dniester. With the addition of a strip of land to the east of Dniester, but with the removal of northern and southern areas, this became the Moldovan Soviet Socialist Republic (SSR) with a territory of some 33,700 km^2.

History and importance

Prior to 1812, Bessarabia was part of the autonomous principality of Moldova under Ottoman suzerainty, but in 1812, it was ceded to the Russian Empire. In 1878, the Congress of Berlin formally recognized the remains of Moldova and Wallachia as the independent Romania, but the whole of Bessarabia was declared to be Russian territory. This caused great dissatisfaction in Romania and led to an irredentist campaign.

Despite subsequent Russification, in December 1917, anti-Bolshevic forces set up a Council which in November of the following year voted for unconditional union with Romania. This was recognized by the signatories of the Treaty of St. Germain, but the transfer was not recognized by the Soviet Union which established the 'autonomous Moldovan' SSR in a strip of Ukranian land on the east bank of the Dniester (October 1924) as a prelude to taking over the whole of Bessarabia. On 26 June 1940, an ultimatum was issued to the government of Romania and Bessarabia was secured for the Soviet Union. Most of the area ceded was added to the Dniester strip, to become the Moldovan SSR, but the southern Black Sea coast area, a territory of 15,000 km^2, was incorporated into Ukraine.

In June 1941, Romania declared war on the Soviet Union and recaptured Bessarabia, which it held until the fall of its government, with the Soviet advance, on 23 August 1944. The subsequent armistice recognized the legality of the Soviet annexation in 1940 and this was later confirmed by the Soviet-Romanian Peace Treaty of 1947.

The Moldovan SSR was particularly important as it supplied 25 per cent of the fruit and vegetables, 23 per cent of the tobacco and 10 per cent of the meat for the Soviet Union. Geopolitically, its significance rests upon the potential for ethnic conflict and the role of the 'Bessarabian Question' in relations between Romania, Russia, Ukraine and Moldova. In the census of 1979, the population was put at just over 3.9 million, 64.6 per cent of whom were Romanian speakers, 14.2 per cent Ukranian and 11.6 per cent Russian. By 1991, the population was estimated to be 3.3 million, comprising 65 per cent Romanian, 14 per cent Ukranian and 13 per cent Russian.

Status

With the collapse of the Soviet Union, the Moldovan Popular Front has taken power in Kishinev and is seeking independence. It has insisted on its own armed forces, within the framework of the Commonwealth of Independent States. A major reason for this is the continuing unrest in the Dniester River valley, where ethnic Russians have formed a 'breakaway Dniester Republic' and have fought with Moldovan Security Forces. Given the continuing unrest in many parts of the Soviet Union, Moldova could well remain a key flashpoint. Should it seek closer relations with Romania, this would disturb the current balance of states in the region.

References

Boundary Bulletin, (1991), No. 1, International Boundaries Research Unit, Durham University.

Day, A.J. (ed.) (1984), *Border and Territorial Disputes*, Longman, London.

The Economist (1991) 6 April.

The Independent (1991) 14 December.

16 Cabinda

Description

Cabinda is an enclave of Angola, 7,270 km² in area, on the Atlantic coast of Africa, just north of the mouth of the Zaïre River. It is separated from the remainder of Angola by some 40 km of coastline, Zaïre's corridor to the sea. The border of Cabinda measures some 425 km, 200 km with Congo to the north, and 225 km with Zaïre to the east and south.

History and importance

The explorer Diego Cao established a trading post at the mouth of the Congo (Zaïre) in the late 15th century, and it is from that time that Portuguese claims to Angola have been made. Control north of the river was not exercised until much later. Portugal occupied Cabinda in 1783 but was ejected by France within 11 months.

Eventually, Portugal laid claim to Cabinda in a convention to the Angola–Portuguese Treaty of 22 January 1815. Much later, the Angola-Portuguese Treaty of 26 February 1884 acknowledged Portuguese claims, including the whole of the Atlantic coast of Africa, between latitudes 5° 12' and 8°S. However, this latter part of the Treaty was never put into effect, following strong protest from other European states. Therefore, Portugal proposed an international conference, which led directly to the Berlin Conference from 15 November 1884 to 26 February 1885.

Portuguese interests north of the Congo were recognized during the Berlin Conference, and on 14 February 1885 Portugal and the International Association of the Congo (later the Belgian Congo, Republic of Congo, Democratic Republic of Congo, Zaïre) signed a treaty delimiting the Cabinda–Congo boundary, recognizing Portuguese sovereignty in the area and guaranteeing the Association a narrow corridor to the sea. France also recognized the Portuguese claims to Cabinda and a Franco–Portuguese Convention of 12 May 1886 established the boundary between French Congo (Congo) and Cabinda.

After the independence of Angola (11 November 1975), Cabinda, which had been stable until that time, had, for a short while, its own insurgency. Both Zaïre and Congo considered the idea of demanding a referendum in 1976, but in the end did not challenge the integrity of Cabinda. It is reasonable to assume that, given Zaïre's potential dependence on the Benguela Railway and the threat of Shaban seccessionist groups operating from Angola, support by Zaïre for Cabindan separatism would be muted.

Although very small, Cabinda has a strategic significance in that it restricts Zaïre's access to the Atlantic. More importantly, offshore oil resources in two fields include 50 billion cubic metres of natural gas and 2.1 billion barrels of petroleum.

Status

Although separatist sentiments remain significant in Cabinda, the issue is at present dormant. However, there is an instability in Zaïre and this may, at some stage, lead to a resurrection of the 'Cabinda Question'.

References

Brownlie, I. (1979), *African Boundaries: A Legal and Diplomatic Encyclopaedia*, Royal Institute of International Affairs, London.

Downing, D. (1980), *An Atlas of Territorial and Border Disputes*, New English Library, London.

Griffiths, I.L.-L. (1985), *An Atlas of African Affairs*, Methuen, London.

Munro, D. and Day, A.J. (1990), *A World Record of Major Conflict Areas*, Arnold, London.

United States Department of State (1970), *Angola-Congo (Brazzaville) Boundary*, International Boundary Study No. 105, 15 October, Office of the Geographer, Bureau of Intelligence and Research, Washington DC.

United States Department of State (1974), *Angola-Zaïre*, International Boundary Study No. 144, 4 April, Office of the Geographer, Bureau of Intelligence and Research, Washington DC.

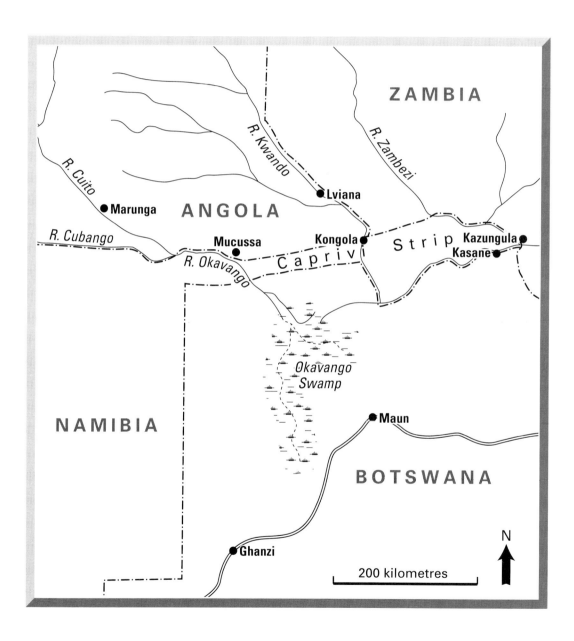

17 The Caprivi Strip

Description

The Caprivi Strip is a narrow panhandle of territory, extending 450 km from the northeast corner of Namibia and giving access to the River Zambezi and Zambia's southern frontier. In contrast to Wakhan (Map 77), where the Panhandle was established largely as a buffer, this is a Panhandle of access.

History and importance

The Caprivi Strip was ceded to Germany by Britain in 1893 to give German South West Africa (Namibia) access to and an outlet through the Zambezi River corridor. It was therefore transferred for strategic reasons, even though the original intent was later found to be impracticable. The Caprivi Strip takes its name from the German foreign minister of the time.

During the Angolan Civil War, the Strip became important as a base for South African intervention against the Soviet-backed Popular Movement for the Liberation of Angola (MPLA). Local Caprivi Baragwena Basarwa were recruited as trackers for the South African Defence Forces (SADF) and formed, under white officers, the 201 (Bushmen) Batallions. Furthermore, the Strip also offered South Africa a route into the heart of the Front Line States from which to launch attacks. For example, the ferry route from Zambia to Botswana, across the Zambezi at Kazungula, the farthest extremity of Caprivi, was regarded by both South Africa and Rhodesia (before it became the majority-rule Zimbabwe) as an infiltration route for insurgents and was attacked from this Strip on several occasions. In March 1990, the last SADF troops left the Caprivi Strip.

Status

Since the independence of Namibia in 1990, there has been no military conflict in the Strip and there are no territorial disputes. However, the Basarwa people, following the withdrawal of SADF financial, medical and financial support, have been left largely destitute. The other problem concerns Zambian smugglers; to deter them the Namibian government established a border post in November 1991 in the Strip, near Manbora. Unless there are further major changes in the neighbouring countries, it is unlikely that the Caprivi Strip will be again a global flashpoint.

References

Boundary Bulletin. No. 1, (1991), International Boundaries Research Unit, Durham University.

Brownlie, I. (1979), *African Boundaries: A Legal and Diplomatic Encyclopaedia*, Royal Institute of International Affairs, London.

Downing, D. (1980), *An Atlas of Territorial and Border Disputes*, New English Library, London.

Munro, D. and Day, A.J. (1990), *A World Record of Major Conflict Areas*, Arnold, London.

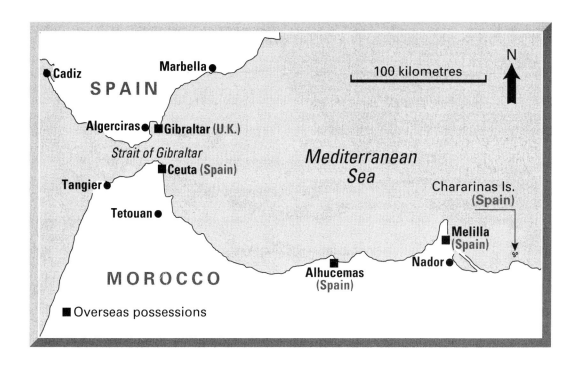

18 Ceuta

Description

Ceuta is a Spanish enclave near Tangiers in Morocco on the opposite side of the strait from Gibraltar. It has an area of 19 km^2, a sea coast 20 km long and a land boundary of 8 km. Its population, predominantly Spanish, is about 67,000. Spain possesses a further series of enclaves in Morocco, the largest being Melilla, located east of Ceuta, near Nador. Melilla has an area of 12 km^2, a sea coast of 3.9 km and a land boundary of 10 km. It too has a predominantly Spanish population of about 65,000. In these two main enclaves there are approximately 19,000 Spanish troops.

Spain also controls the Rock of Velez (Peñon de Velez de la Gomera), the Rock of Alhucema (Penoes de Alhucemas), both small groups of islands just off the Moroccan coast between Ceuta and Melilla, together with the Chafarinas Islands, which lie 43 km east of Melilla. The Penones have a civilian and military population of about 100 each and the Chafarinas a population of approximately 200.

On the other side of the strait is Gibraltar, a British controlled enclave in mainland Spain. It was developed as a military base and port city and has an area of 5.8 km^2 and a population of about 30,000. For some 250 years, Gibraltar was one of Britain's most valuable naval bases.

The Strait of Gibraltar is approximately 58 km long and 12.5 km at its narrowest point. In the main shipping channels, depths vary from 935 to 320 m. In terms of international shipping, it is the second busiest international strait after the Straits of Dover. (In sheer number of vessels of all sizes, only the Strait of Malacca is more important.) More than 150 vessels, each of over 1,000 gross tonnes, transit daily, making an annual total of almost 55,000 ships. Approximately one-third of these are oil tankers and, with the increased capacity of both the Suez Canal and the SUMED pipeline, there has been significant growth in the oil trade.

Therefore, for the West, Gibraltar has replaced Hormuz as the key choke point, since the SUMED pipeline means that the latter may now be by-passed.

History and importance

With regard to the Spanish enclaves, dates of occupation or annexation were: Ceuta 1580, Melilla 1497, the Rock of Alhucema 1673, the Rock of Valez 1508 and the Chafarinas 1848. Gibraltar was occupied by the Moors between 711 and 1462 and was then held by the Spanish until 1704. It was captured by the Anglo–Dutch fleet on 4 August 1704 and acquired by Britain following the Treaty of Utrecht on 17 July 1713. Thus, Britain has held Gibraltar longer than Spain. Since independence in 1956, Morocco has pressed for the return of Ceuta and the other enclaves, while Spain has exerted pressure on Britain for the return of Gibraltar since the end of World War II, though the force of the claim has lessened following Spain's return to democracy and both countries' membership of the EC.

In 1961, Morocco called on the United Nations to recognize its rights with regard to Ceuta and the other enclaves, but the result was that Spain merely reinforced their borders. On 29 June 1962, Morocco reasserted its claim and the next day extended its territorial waters from 6 to 10 nml. Between 1963 and 1965 meetings between King Hassan and General Franco eased tensions. When Gibraltar was given fuller internal self-government in 1964, Spain limited trade with the enclave, obstructed frontier crossings and threatened to impede communications.

From 1969 onwards, land communications were stopped completely, to be reopened in 1982. On 14 June 1967, in a referendum, 99 per cent of Gibraltarians were in favour of retaining the British link and, in each subsequent election for their House of Assembly they reaffirmed this decision. On 27 January 1975, Morocco formally requested the UN

Decolonization Committee to put the Ceuta case on its agenda, and between 1978 and 1979 the Moroccan Patriotic Front carried out several bomb attacks in Ceuta and Melilla. Morocco has always stated that it would raise the question of the two 'presidios', as both Ceuta and Melilla are known in Morocco, and other enclaves if Gibraltar were to be transferred to Spanish control.

The enclaves on both sides of the strait are important militarily and strategically, since, by almost any criterion, the Strait of Gibraltar is vital. Furthermore, the ownership of the enclaves entitles the countries concerned, under the terms of UNCLOS (1982), to a large share of the Mediterranean sea-bed. With Ceuta and the small, strategically placed island of Alboran, Spain is in a position virtually to close off entry to the Mediterranean.

Status

In October 1991, Spain categorically rejected the idea that Gibraltar could become a self-governing dependency of the EC. However, as King Hassan has stated, if Spain regains Gibraltar, Morocco would retrieve the enclaves and islands, since 'no power can permit Spain to possess both keys to the same straits'. Over most of the post-war period, conflict has been focused upon the eastern end of the Mediterranean Sea, but there is an increasing possibility, particularly in the light of the problems in Algeria, that the western end may become the more important flashpoint.

References

Blake, G.H. and Drysdale, A. (1985), *The Middle East and North Africa: A Political Geography*, Oxford University Press, Oxford.

Blake, G.H., Dewdney, J. and Mitchell, J. (1987), *The Cambridge Atlas of the Middle East and North Africa*, Cambridge University Press, Cambridge.

Boyd, A. (1991), *An Atlas of World Affairs*, Routledge, London.

Day, A.J. (ed.) (1984), *Border and Territorial Disputes*, Longman, London.

The Economist Atlas (1989), Economist Books, Hutchinson, London.

The Observer (1990), Gibraltar: Rock seeks home rule within Europe, 2 December, p. 14.

Prescott, J.R.V. (1985), *The Maritime Political Boundaries of the World*, Methuen, London.

19 The Chagos Archipelago (Diego Garcia)

Description

The Chagos Archipelago consists of six major islands, of which the largest, an atoll some 21 km long and 6 km wide, is Diego Garcia (6° 34′S, 71° 24′E). It is located approximately 1,000 nml south southwest of India, 2,000 nml southeast of the Persian/Arabian Gulf and 1,200 nml northeast of Mauritius. When in 1965 the island became a joint United States–Great Britain base, over 1,000 people were resettled from Diego Garcia to Mauritius. There were 508 families forcibly removed in the period 1965–71 and the remainder suffered a mass evacuation between 1971 and 1973. A comprehensive survey of the Chagos refugees in Mauritius in 1981 put their number at some 2,800.

History and importance

Discovered by the Portuguese in 1532 and the scene of an abortive British attempt to establish a supply station in 1786, Diego Garcia came under French control until the end of the Napoleonic Wars when it passed to Britain. During World War II, it was important for ship repairs and refuelling. In the late 1950s, the United States, which needed to secure for its Navy a replenishment base and airfield in the world's third largest ocean, felt the need to use the islands for this purpose. On 8 November 1965, the British Indian Ocean Territory (BIOT) was established, consisting of the Chagos Archipelago, formerly administered with Mauritius, and the Seychelles Islands of Aldabra, Farquhar and Desroches. These three islands were returned to the Seychelles upon that country's independence on 29 June 1976.

In December 1966, the United States and Britain signed a defence agreement, leasing the BIOT to the United States for 50 years, with an option for an extra 20 years. There followed a steady development of facilities, including the construction of a 4,000 m runway, which allowed the airfield to accommodate B52 bombers. These developments gained added impetus following the Iranian revolution (1979) and the loss of facilities at Bandar Abbas and Chah Bahar and the Soviet invasion of Afghanistan. Indeed, on 4 January 1980, the United States announced that it had decided to reinforce facilities on Diego Garcia and to maintain a permanent presence of up to 4,500 troops. Between 1981 and 1986, over $1 million was invested, allowing the prepositioning of seven, later 17, vessels with equipment for a marine amphibious brigade of 12,000 men.

Since it is remote, secure and centrally placed, Diego Garcia remains vital for American military, particularly naval, interests in the Indian Ocean. It is relatively immune from land-based air attack and also, since it has been depopulated, from local political conflict. It provides a deep anchorage, refuelling facilities, a repair capability and a well-appointed airbase. It can support a full carrier task force and has prepositioned equipment, communication facilities and a surveillance capability. It is not a full-sized base on the scale of Subic Bay and is a long way from any mainland, particularly the Gulf area. Furthermore, for a variety of reasons, it provides a focus for anti-American feeling in the region.

Status

Despite the changing global power distribution, Diego Garcia is likely to remain a key strategic location for the foreseeable future. Given the problems in almost all of the littoral states, and the current role of the United States as the world's policeman, the atoll provides an ideal base. Furthermore, as was illustrated during Operation Desert Storm, it facilitates rapid deployment crucial to the Persian/Arabian Gulf region.

The major problems concern local sensitivity, and in June 1980 all the Mauritian political parties called for the return of Diego Garcia. On July 4

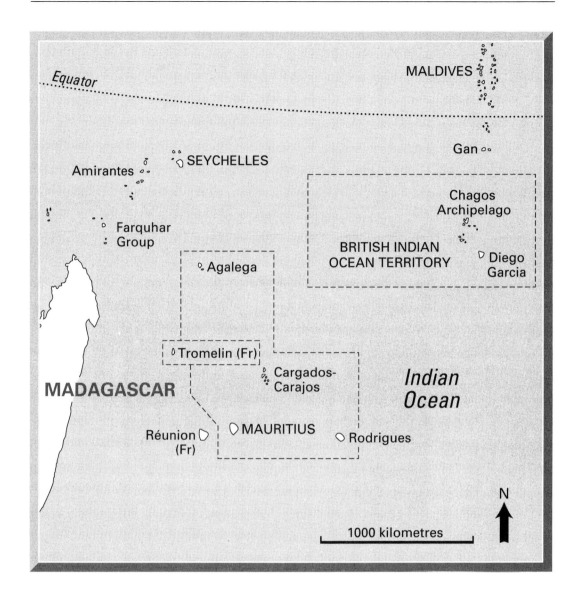

1980, the Organization of African Unity (OAU) heads of state called for the return of the atoll and its demilitarization. The High Commissioner of Mauritius stated on 7 April 1974 that the United States' use of Diego Garcia as a military base would violate the various British undertakings given. The British government denied this and the dispute continues, as does the grievance of those forcibly resettled on Mauritius, many of whom have so far refused financial compensation. Diego Garcia therefore remains an example of a classic global flashpoint which could be the focus of conflict on a global, regional or local scale.

References

Anderson, E.W. (1988), *Strategic Minerals: The Geopolitical Problem for the United States*, Praeger, New York.

Cottrell, A.J., and Hahn, W.F. (1978), *Naval Race or Arms Control in the Indian Ocean*, Agenda Paper No. 8, National Strategy Information Center, Inc., New York.

Day, A.J. (ed.) (1984), *Border and Territorial Disputes*, Longman, London.

Downing, D. (1980), *An Atlas of Territorial and Border Disputes*, New English Library, London.

Prescott, J.R.V. (1985), *The Maritime Political Boundaries of the World*, Methuen, London.

Rais, R.B. (1986), *The Indian Ocean and the Superpowers*, Croom Helm, London.

20 The Curzon Line

Description and history

Following the victory of the Bolsheviks in the Russian Civil War, the repulse of the first of the Polish invasions of Russia and the Soviet invasion of Poland, the Western Powers proposed at the Spa Diplomatic Conference in Belgium (July 1920) a mediated frontier along the Curzon Line. In October 1920, the frontier was finally settled by the Treaty of Riga, giving Poland a large White Russian and Ukranian minority.

In the Yalta Agreement (4–10 February 1945), the three heads of the allied governments (Churchill, Roosevelt and Stalin) agreed that the eastern frontier of Poland should follow the Curzon Line, with certain digressions of between 5 and 8 km from it in favour of Poland. Thus, despite the changes during World War II, the Curzon Line retained its long-term importance. Subsequently, it was confirmed as being inviolable by the Final Act of the Conference on Security and Cooperation in Europe, signed in Helsinki on 1 August 1975 by 35 countries, including every European state except Albania.

Importance

The Curzon Line is of historical importance as the marker of Poland's eastern boundary. Its continuation southwards enabled the Soviet Union to adapt the boundaries of Eastern Europe for its own strategic and economic purposes.

Status

The eastern boundary of Poland is, in some respects, the temporary eastern boundary of Europe. Eastwards, the breakdown of the Soviet Union has resulted in a variety of problems of possible realignments. The Polish frontier is the boundary of stable Europe. Furthermore, it may well become the eastern boundary of NATO.

The fundamental problem of the region is the potential flood of refugees from the former Soviet Union, and on 6 March 1991 the Polish national defence minister stated that a major redeployment of the Polish army from western to eastern Poland had begun. The plan for the strengthening of Poland's border with the former Soviet Union (CIS) was reported to have been in preparation for six months and was prompted by the anticipated uncontrolled migration from Russia and the other successor states. Indeed, in November 1990, it was reported that there were already some 70,000 Soviet citizens in Poland, together with approximately two million ethnic Poles in the Soviet Union who could claim refugee status. Therefore, for entirely different reasons to those that obtained during the Cold War, the Curzon Line may yet again become a flashpoint.

References

Boundary Bulletin, No. 1, (1991), International Boundaries Research Unit, Durham University.
Boundary Bulletin, No. 2, (1991), International Boundaries Research Unit, Durham University.
Day, A.J. (ed.) (1984), *Border and Territorial Disputes*, Longman, London.

21 East Timor

Description

Timor is a large island, some 33,900 km² in area, lying to the north of Australia and east of Indonesia, with which part of it has a disputed relationship. The estimated population (1991) is 650,000 and average per-capita annual income is $200, some 36 per cent of the Indonesian average.

History and importance

The Portuguese established themselves on Timor in 1520 and in 1680, Western Timor was taken over by the Dutch. This division, established so early, has resulted in continuing conflict to the present day. East Timor became a Portuguese colony in 1702 and a self-governing colony in 1926.

In 1949, West Timor joined what had been the Dutch East Indies to become part of the Republic of Indonesia. East Timor remained a Portuguese colony until the fall of the Caetano government in Portugal in April 1974, when the Presidents of Indonesia and the government of Australia agreed that the best interests of the Timorese lay in Indonesian annexation. In May of that year, the Portuguese minister for inter-territorial coordination promised a referendum on the decolonization process, but this failed to take place.

During 1975, several laws were passed concerning Timorese self-determination, but on 21 August Portugal lost control and a full-scale civil war between the Revolutionary Front for Independence (FRETELIN), and the other movements and parties began. By 8 September, FRETELIN claimed complete control of East Timor and a memorandum of understanding between Portugal and Indonesia, agreed on 3 November, confirmed Portugal as the legitimate authority. On 28 November, FRETELIN announced the independence of the Democratic Republic of East Timor, a move which the pro-Indonesian parties claimed removed the last remains of Portuguese sovereignty, thus legitimizing the union with Indonesia. Portugal requested United Nations' assistance, but on 7 December, 1,000 Indonesian troops entered East Timor and Portugal broke off diplomatic relations. By the end of the year, various parts of East Timor, including the enclave of Oekusi Ambeno had been annexed.

The United Nations condemned Indonesia, but on 21 March 1976 the Indonesian foreign ministry announced the formation of a parliament in East Timor to sanction integration. In August that year, East Timor was declared Indonesia's 27th province to be known as Loro Sae. The operation had cost 100,000 lives and a further 300,000 people were incarcerated, but the Western world took little notice, since, in the wake of Communist victories in Vietnam and Cambodia, Indonesia was considered as a key bulwark against Communism whose internal affairs were best left unnoticed.

In January 1978, the government of Australia recognized the annexation, but at a 1979 conference the Non-aligned Movement adopted a resolution reaffirming the right of the Timorese to self-determination. In 1984, Indonesia attracted criticism from both the United States and the Vatican concerned for the Catholic minority, for its violation of human rights; and the Amnesty International report of June 1985 stated that some half-million people had been killed or resettled since the 1975 invasion. From 1986 onwards, there have been reports of FRETELIN activity, but the guerrillas number at the most 1,500 and are opposed by some 60,000 Indonesian troops.

Status

East Timor is important, not only as a battleground for self-determination, but also because the Timor Sea has large proved oil reserves. This factor has also resulted in the muted Australian response to Indonesian atrocities. In 1989, Australia and Indonesia agreed a zone of cooperation in the Timor Sea which, since, for various reasons, it does not use the median-line principle, may well be a model for other such cases.

Meanwhile, the insurgency in East Timor continues and on 12 November 1991, between 50 and 60 people were killed when Indonesian troops opened fire on a peaceful demonstration at Santa Cruz cemetery. This put Australia and Japan in particular under pressure to cut economic aid to Indonesia. However, Indonesia is an important source of hydrocarbons for Japan, and Australia has no wish to disturb a relationship which is allowing it to go ahead with oil exploration in the Timor Sea. Thus, East Timor remains an active flashpoint, despite the efforts of most influential foreign countries to play down the issue.

References

Boundary Bulletin, No. 1, (1991), International Boundaries Research Unit, Durham University.

Boundary Bulletin, No. 3, (1992), International Boundaries Research Unit, Durham University, January.

Boyd, A. (1991), *An Atlas of World Affairs*, Routledge, London.

Day, A.J. (ed.) (1984), *Border and Territorial Disputes*, Longman, London.

Downing, D. (1980), *An Atlas of Territorial and Border Disputes*, New English Library, London.

The Economist Atlas (1989), Economist Books, Hutchinson.

The Economist (1991), 12 January.

The Guardian (1991), 19 November.

The Independent (1991), 'Massacre among the Graves', 17 November.

The Independent (1991), 19 November.

The Independent (1991), 22 November.

The Independent (1991), 24 November.

Munro, D. and Day, A.J. (1990), *A World Record of Major Conflict Areas*, Arnold, London.

22 Epirus

Description

Epirus (or Ipiros) is the rugged region that straddles the Greek–Albanian border. The shared border is approximately 280 km long. In 1939, the Greek population in southern Albania, northern Epirus, was estimated to be some 300,000 or 20 per cent of the total population of Albania. After World War II, many Greeks left Albania and in 1981 the estimated Greek population was 200,000. However, even as late as the 1980s, 62 of Albania's 250 members of parliament were of Greek extraction. The history of Epirus reflects the very complex political history and geography of the region.

History and importance

On 12 November 1912, during the first Balkan War, Albania proclaimed its independence and in December this was recognized by the European powers at the Peace Conference in London. During the summer of the following year, the ambassadors of the Great Powers agreed in principle to the borders at a conference in London. However, in October 1914 Greece occupied southern Albania and from 1915 Austro–Hungarian forces occupied the north and centre.

In a secret 1915 treaty to bring Italy into the war on the Allied side, Greece was promised the south, Serbia the north and Italy, central Albania. In 1920, the Albanians again declared their own independence and ejected the Italians who withdrew in August, recognizing the independence and territorial integrity of Albania. Albania was then admitted to the League of Nations. By 1926, the British, Italian and French Boundary Commission had completed its work and the final Demarcation Act was signed by Greece and Yugoslavia on 30 July 1926 in Paris. This was followed in November by a Treaty of Friendship and Security between Italy and Albania.

On 17 April 1939, Albania was occupied by Italian troops, and in October 1940 Italy attacked Greece from Albania. However, the Italians were repulsed and Greece occupied about half of Albania. Then, in April 1941, German forces overran Yugoslavia and Greece but were resisted in Albania by a National Front of Communists and Nationalists.

In 1945, the National Front was recognized by the Allies, but on 10 November of that year, the Greek government protested, demanding 'the union of north Epirus with the Greek Motherland'. In July of 1946, the 'Pepper Resolution', favouring the ceding of north Epirus to Greece, was passed by the United States Senate. But after protests from Albanian leader Envar Hoxha, in August and September 1946, the Epirus issue was struck off the agenda of the Paris meeting of Allied foreign ministers, by the American secretary of state.

Thus, the Allies gave *de facto* recognition to the 1913 frontiers. On 2 July 1958, the Albanian government rejected the Greek statement that a state of war still existed between the two countries and asked for normal relations. Greece reiterated its claim, and on 14 August this was rejected by Albania in a statement which affirmed that no state of war existed and 'the question of north Epirus does not exist, as this is Albanian territory'. In 1962, the Albanian government tried again for diplomatic relations, but these were not achieved until 6 May 1971, when a peace treaty was signed, which implied recognition of Albanian borders. In 1981, the Greek prime minister stated that Greece was opposed to 'any attempt to disturb the status quo in the area'.

Status

The north Epirus issue is now, to all intents and purposes, dead, but, despite *rapprochement* from the 1970s onwards, the issue has remained an underlying factor in Greek–Albanian relations. With the current economic and political chaos in Albania, it is possible that the 200,000 ethnic Greeks in Albania

may wish to leave or to seek Greek protection. Equally, Albanian refugees of non-Greek ethnicity may attempt to enter Greece.

One further issue concerns the Corfu Channel. Albania has not negotiated any maritime boundaries with its neighbours, but is very likely, when it does, to seek redress for the constricting effect of Corfu (Kerika), Erikousa and Othomoi. An Albanian–Greek offshore boundary dispute would involve many of the issues already extant in the Greece–Turkey case.

References

Day, A.J. (ed.) (1984), *Border and Territorial Disputes*, Longman, London.

Prescott, J.R.V. (1985), *The Maritime Political Boundaries of the World*, Methuen, London.

United States Department of State (1971), *Albania–Greece*, International Boundary Study No. 113, 18 August, Office of the Geographer, Bureau of Intelligence and Research, Washington DC.

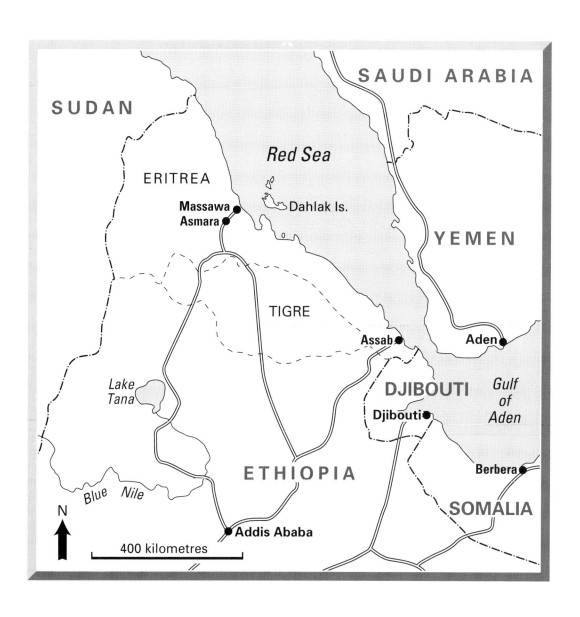

23 Eritrea

Description

Eritrea is a region of Ethiopia. It runs along the Red Sea from Sudan in the north to Djibouti in the south. Formerly an integral part of Ethiopia, it comprises all of Ethiopia's sea coast, including the crucial ports of Assab and Massawa. In the south, it is under 100 km wide, but in the north it broadens to include the high plateau extension of the Ethiopian Highlands and the western lowland area bordering Sudan. Eritrea has an area of 117,600 km^2 and a population, estimated in 1984, to be 2.6 million. With regard to ethnicity and religion, the area is very mixed.

Approximately half of the population comprises Tigrinya speakers on the plateau who are mainly Christian and share their language and religion with the neighbouring Tigre province. The same language is spoken among the Muslims of the western lowlands and northern coastal strip who comprise about one-third of the population. In the southern coastal strip are the Danakils, Muslim nomads. Other small language groups, Christian and Muslim, inhabit enclaves in the north. The coastline extends for 1,070 km from Bab el Mandeb (q.v.) to Ras Kasar on the Sudanese border and includes the two key ports of Massawa and Assab.

History and importance

Italian influence has long been strong in the region and in the 1880s Italy established colonies at both Assab and Massawa. By 1890, it controlled the whole coast from Sudan (British) to French Somaliland (now Djibouti). In 1896, a protectorate over Ethiopia was declared, but the Italians were defeated at the battle of Adowa and it was not until 1936 that they were able to control the country. From 1936 until 1941 Ethiopia and Eritrea were ruled together as part of the Italian East African Empire. In 1941, Eritrea was placed under British military administration which lasted for nine years. Meanwhile, a Four-Powers Commission failed to agree on the future for Eritrea. The alternatives were: union with Ethiopia (suggested by Britain), partition (the United States), trust territory with Italian administration (France) and trust territory with international administration (the Soviet Union). The United Nations Commission favoured close association with Ethiopia, and UN Resolution 390A to that effect was passed in December 1950.

In September 1951, Eritrea became an autonomous territory federated with Ethiopia. Within Eritrea, Ethiopia financed unionists and terrorized pro-independence movements, but, as a potentially strong ally in the Red Sea, its government gained American support. This was given expression in 1953 by a Defence Pact between Ethiopia and the United States. Ethiopia then continued to pursue union with Eritrea and on 14 November 1962 unilaterally imposed direct control. This led to the formation of the Eritrean Liberation Front and a 30-year civil war.

In 1974, as a result of the Ethiopian revolution, Emperor Haile Selassi was overthrown and a neo-Marxist government installed. In December 1976, the new government signed a military assistance pact with the Soviet Union. In 1977, with the commencement of the Ogaden War, the Eritrean independence movements were able to gain control of much of Eritrea, outside the garrison towns, but in 1978, the Ethiopian Army, backed by some 15,000 Cuban troops, recaptured most of the rebel-controlled territory.

During the period 1983-5, severe drought and famine afflicted the country and the Ethiopian government directed international aid away from Eritrea, but by 1987 the Eritrean Peoples' Liberation Front (EPLF) had gained effective control of northern Eritrea and the port of Massawa was cut off. By 1988, the Ethiopian army had been largely defeated in Eritrea, and Ethiopia was cut off from the Red Sea coast. In May 1991, the Marxist regime of Mengistu collapsed and the coalition government that came to power acknowledged Eritrea's right to secede and form its own independent country.

Eritrea is of course vital because it controls Ethiopia's access to the sea. It has been successively important in American and Soviet naval strategy and the Soviets, in particular, constructed major bases at Massawa and in the Dahlak archipelago. Furthermore, Eritrea has a strategic location, being adjacent to the vital choke point of Bab el Mandeb.

Status

Although it has achieved effective independence, and set a precedent for other insurgent groups, Eritrea remains extremely poor. Its economy is under-developed and suffering severely from war damage, together with the effects of drought and famine. Approximately 170,000 refugees fled into Sudan and these may well wish to return. To the south, the Damakil are related to the Afars of Djibouti and boundary issues seem a possibility. Internally, Eritrea's mixed ethnic and religious mosaic could provide the fuel for further conflict. But of greater concern is the fact that it seems unlikely that, in the long term, Ethiopia will accept the present position, and therefore Eritrea is likely to remain a potential flashpoint.

References

Anderson, E.W. (1991), 'Making Waves on the Nile', *Geographical Magazine*, **LXIII**, No. IV, pp. 10–13.

Boundary Bulletin, No. 1, (1991), International Boundaries Research Unit, Durham University.

Boundary Bulletin, No. 2, (1991), International Boundaries Research Unit, Durham University.

Boundary Bulletin (1992), No. 3, International Boundaries Research Unit, Durham University, January.

The Economist (1991) 1 June.

Griffiths, I.L.-L. (1985), *An Atlas of African Affairs*, Methuen, London.

Munro, D. and Day, A.J., (1990), *A World Record of Major Conflict Areas*, Arnold, London.

24 The Falkland Islands (Malvinas)

Description

Located in the far south of the South Atlantic (Port Stanley is at 51° 45'S, 57° 56'W), the Falkland Islands are situated some 480 km east of Argentina. There are two main and about 200 smaller islands, with a total area of 11,950 km². The population of 1,958 (1990) only marginally exceeds the number of British troops (1,600 in 1989), guarding the islands. South Georgia, a dependency of the Falkland Islands involved in an early stage of the Falklands War (1982), has an area of 3,750 km² and lies 1,285 km east-southeast of the Falklands.

History and importance

There are British records from the previous century, but the first fully authenticated sighting of the Falklands was made by the Dutch in 1600 and the first landing by the British in 1690. In 1764, the French claimed East Falkland and established a settlement, Port Louis. The Spanish name for the islands (Malvinas) derives from the French 'Malouines' from the name of the home port of the French fishermen, St. Malo. One year later, in ignorance of the French position, the British landed on West Falkland, established a settlement, Port Egmont, and claimed the entire island group. When the French were discovered, they were invited to leave and sold their settlement to Spain, by whom it was named Puerto de la Soledad. It was placed under the jurisdiction of Buenos Aires, and in 1769, the British asked the Spanish to leave. The response was a military force from Buenos Aires, the departure of the British and very nearly a war between Spain and Britain.

In 1771, there was an Exchange of Declarations and both sides accepted the return to the status quo with Port Egmont being returned to Britain. However, in many ways the most crucial event in the history of the Falklands occurred in 1774 when Britain withdrew its garrison from Port Egmont, leaving merely a British flag and a plaque recording British ownership. The British departure occurred as a result of economic considerations and, for the same reason, Spain also abandoned the Falkland Islands in 1811.

In 1816, the province of Buenos Aires declared its independence from Spain and in 1820, a Falklands claim was lodged on its behalf. A governor was appointed to the Falklands by the Buenos Aires government in 1823, in which year the United Provinces of La Plata were recognized by the United States. Britain also recognized the government one year later and in 1825 signed a Treaty of Amity, Trade and Navigation. No mention was made of the Falkland Islands and, in June 1829, the Buenos Aires government claimed sovereignty over 'the Islands of the Malvinas and those adjacent to Cape Horn in the Atlantic Ocean'. Britain immediately protested, basing its case on the fact that its withdrawal in 1774 did not terminate its claim. The United States supported the British claim and later heated exchanges followed between Argentina on the one hand and Britain and the United States on the other.

Eventually, between 1832 and 1833, Britain sent a warship and took possession of both West and East Falkland. Settlers from Britain arrived both to develop sheep farming and to provide a port of refuge for vessels on the Cape Horn route, and the Falkland Islands Company, a British-based trading company, was established. In 1908, Britain extended its sovereignty to South Georgia and the South Sandwich Islands. Later, both were claimed by Argentina and the case was submitted by Britain to the International Court of Justice. However, Argentina and Chile, both disputing the British claims, did not recognize the jurisdiction of the Court and therefore no action could be taken. In 1966, the United Nations initiated talks between Britain and Argentina over the Falkland Islands and in 1972, an air link was established between

The Falkland Islands (Malvinas)

them and Comodoro Rivadavia in southern Argentina.

Talks dragged on in a desultory fashion until, in 1980, Britain offered the Legislative Council of the Falklands a number of options:

(a) a 25-year freeze on the dispute;
(b) the transfer of sovereignty to Argentina, with a lease-back by Britain and
(c) a joint British–Argentinian administration.

The Legislative Council expressed a preference for the first option, but this was rejected by Argentina. As a result, in February 1982, as Britain was considering withdrawing its one naval patrol ship, HMS *Endurance*, from the area, the government of Argentina sent a group, described as 'scrap-metal dealers', to South Georgia and raised the national flag there. On 2 April, Argentinian troops landed on East Falkland, overwhelmed the small garrison of Royal Marines, and an Argentinian governor was appointed.

Despite a United Nations Security Council Resolution, Argentina refused to withdraw and Britain dispatched a large naval Task Force. On 25 April, South Georgia was retaken and on 14 June, the Argentinian forces in the Falklands surrendered. However, the short war cost 1,000 lives, and the loss of the Argentinian battle-cruiser, the *General Belgrano* and a number of British warships, including the destroyer *HMS Sheffield*. On 22 July the 200 nml Total Exclusion Zone, established during the war, was replaced by a 150 nml Protection Zone. Before the end of the year, Britain had established the Falkland Islands Development Agency and had agreed to provide, over six years, a sum of £31 million for development. Central to this was the construction at Mount Pleasant of a military airport which was completed in 1985.

Further disagreements followed, particularly over Exclusive Economic Zones and fisheries and it was not until October 1989 that the formal end to hostilities was declared. On 15 February 1990, full diplomatic relations were established between Argentina and Britain.

Until 1982, the Falklands merely represented the last vestiges of the British Empire and were of little significance to Britain. On the other hand, geopolitically and emotionally, they had always been crucial to the Argentinians. However, in 1982 the issue of self-determination evoked great public sympathy in Britain and the Falklands attained a high profile. Additionally, with further explorations of the offshore resources, the significance of the Falklands in controlling large areas of fishing and seabed hydrocarbon deposits has been realized. Furthermore, the possession of the Falklands might possibly affect Britain's future claims to Antarctica. Finally, there is the strategic aspect of Cape Horn and its neighbouring inter-oceanic passages, which might become more important choke points.

Status

In 1990, joint negotiations were held in Rio and Madrid on fishing rights and a Joint Statement on Conservation of Fisheries was agreed on 28 November. Further cooperation continued into 1991, when an agreement was reached on banning illegal fishing (this was aimed primarily at the Taiwanese). However, on 22 November, Britain, in the interests of oil exploration, asserted jurisdiction 'to the full limits allowed under international law', while Argentina promulgated a law asserting sovereignty over the Continental Shelf of the Falklands. Nevertheless, under their joint sovereignty 'umbrella', which protects their individual claims, Argentina and Britain continue to reach agreement and to expand cooperation in practical matters. It is likely to be at least 10 years before the effective exploitation of oil is possible, and in that period political and legal accord may be reached.

References

Boundary Bulletin, No. 1, (1991), International Boundaries Research Unit, Durham University.
Boundary Bulletin, No. 2, (1991), International Boundaries Research Unit, Durham University.
Beck, P. (1991), 'Fisheries Conservation: A Basis for a Special Anglo-Argentine Relationship' in *Boundary Bulletin*, No. 2, International Boundaries Research Unit, Durham University, pp. 29–36.
Blake, G.H. and Anderson, E.W. (1982), *International Maritime Boundary Delimitation and Islands: Examples Analogous to Libya–Malta*, University of Durham (unpublished), September.
Boundary Bulletin, No. 1, (1991), International Boundaries Research Unit, Durham University.
Boyd, A. (1991), *An Atlas of World Affairs*, Routledge, London.
Child, J. (1985), *Geopolitics and Conflict in South America*, Praeger/Hoover Institution Press, Stanford.
Day, A.J. (ed.) (1984), *Border and Territorial Disputes*, Longman, London.
The Independent (1991), 23 November.
Morris, M.A. (1986), 'E.E.Z. Policy in South America's

Southern Cone' in E.M. Borgese and N. Ginsburg (eds), *Ocean Yearbook 6*, University of Chicago Press, Chicago, pp. 417-37.

Munro, D. and Day, A.J. (1990), *A World Record of Major Conflict Areas*, Arnold, London.

25 The Gaza Strip, the Golan Heights and the West Bank

Description

The Gaza Strip comprises a narrow belt of coastal zone, controlling approaches from the north to the Sinai Peninsula and from the south to Israel. With an area of only 363 km^2 and an estimated population of 750,000 (1991), the Gaza Strip is extremely densely populated. The town of Gaza has a population of 120,000 and many of the remaining Palestinians inhabit long-standing refugee camps. The government of Israel has established a number of settlements in the Strip and maintains strict military surveillance over it.

The Golan Heights, a section of plateau and escarpment 50 km long, which overlooks Israel, is Israeli-occupied Syrian territory. It extends between Mount Hermon and the mouth of the Yarmuk River and reaches a height of 1,204 m. Having been the scene of almost continous local conflict, it was targeted by the Israeli Army in 1967, was captured and has been retained ever since. The Arab population, numbering some 90,000, fled into Syria, but approximately 12,000 Druz remained.

Since 1949, some 5,880 km^2 of territory to the west of the river Jordan, known as the West Bank, was under the jurisdiction of Jordan. In 1967, it was captured by Israel and remains the largest and most contentious of the Occupied Territories. The population of the West Bank, comprising 97 per cent Palestinian Arabs, totals 866,000 (1990).

History and importance

Since the fundamentals of the history of all three areas are similar, they will be examined together. In the years 70 and 135 AD, Jerusalem was destroyed and thus began the Jewish diaspora. In 636, the city was captured by the Arabs and, despite the crusades of the 11th, 12th and 13th centuries, it remained part of the Muslim world.

Meanwhile, during the late Middle Ages, the Jews began to return and this migration was intensified in the latter part of the 19th and earlier part of the 20th centuries as a result of pogroms against the Jews in Russia and Poland. By 1914, it is estimated that the Jewish population of Palestine had reached, perhaps 90,000, while the Arab population stood at approximately 450,000. As Jewish numbers increased, so conflict with the Arabs grew, particularly as a result of the concept of 'Zionism', promulgated by Theodor Herzl. The basic tenet of this movement was the establishment of a home for the Jews, and money provided by the Jews of Western Europe helped purchase land to this end. The first kibbutz was established at Degania in 1908 and Tel Aviv was built as a Jewish city.

During World War I, Britain and France constructed plans to dismember the Ottoman Empire between themselves but also to create an independent Arab state. However, in 1917, in courting Jewish support for the war, Arthur Balfour, the British foreign secretary, publicly supported the idea of a Jewish homeland in Palestine, the Zionist aim. In December of that year, Jerusalem was captured by the British forces. At the Treaty of Sèvres, on 10 August 1920, the British and French effectively enacted the Sykes–Picot Agreement, whereby France established a mandate over Syria and Britain a mandate over Palestine and Mesopotamia. Despite strong anti-Jewish feeling among the Arabs and a keen sense of betrayal, Britain was forced by the League of Nations to press ahead with the Balfour Declaration and, thus, Jewish national identity was officially recognized. In 1923, Transjordan was established as an autonomous state and the Golan Heights were transferred from the British to the French mandate.

By the latter part of the 1920s, Britain was faced with the appalling dilemma of facilitating the establishment of a Jewish state while, at the same time, in the interests of restricting Arab–Jewish conflict, limiting Jewish immigration. As a result, Britain received increasingly less cooperation from both

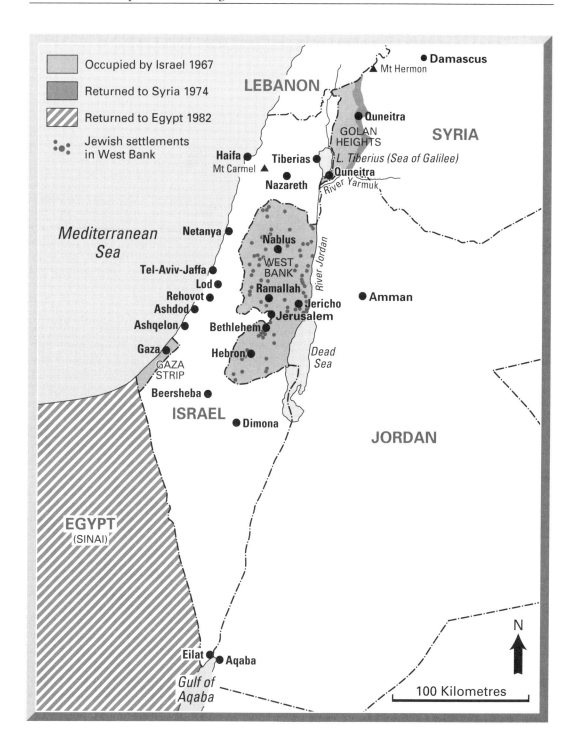

Arabs and Jews. In 1936, with the persecution of Jews in Nazi Germany, the number of immigrants surged so that, by then, the Jewish population of Palestine numbered some 400,000 or approximately 30 per cent of the total.

As a possible solution, in 1937, Britain suggested the partition of Palestine into separate Arab and Jewish states, but, while this idea was reluctantly accepted by the Jews, it was totally rejected by the Arabs. During World War II, inter-communal violence was largely suspended, but the effect of the holocaust merely added to Jewish support for the Zionist cause. In 1947, the United Nations Special Committee on Palestine recommended that Palestine should be partitioned into Jewish and Arab states, a solution which met with a similar response to the earlier suggestion promulgated by Britain. In December of that year, Britain declared that it would terminate its mandate over Palestine on 15 May 1948.

On 14 May, the Jewish National Council declared the New State of Israel, which was almost immediately invaded by Arab forces from Iraq, Egypt. Transjordan, Syria and Lebanon. These attacks were repulsed, and the new Jewish state gained some 15 per cent more territory as a result of the fighting. As a consequence of the Armistices of 1949, Israel gave up control of southern Lebanon, northern Sinai and the Gaza Strip but retained control over more than 60 per cent of the former territory of Palestine. Within the boundaries of this state of Israel, large numbers of Arabs were displaced but some 15,000 remained. Many of those displaced moved to refugee camps, established in the West Bank and the Gaza Strip.

These refugee camps have remained effectively pawns in the Arab–Jewish struggle ever since. They serve as a reminder of Israeli military action and, to retain a claim to what is seen as their rightful home in Palestine, the neighbouring Arab states refuse to integrate them. In April 1950, the West Bank and East Jerusalem were incorporated into Jordan. In 1956, in collusion with Britain and France, Israel entered the Sinai Peninsula and Gaza Strip during the Suez Canal conflict. As a result of pressure from both superpowers, Israel was forced to withdraw and the United Nations established an Emergency Force in both areas. Arab frustration deepened and in 1964 the Palestinian National Charter was adopted and the Palestine Liberation Organization (PLO) established as the principal anti-Zionist movement.

One year later, the Fatah guerrilla organization received official approval under a joint leadership which included Yasser Arafat. Following further fighting, in 1967, President Nasser closed the Strait of Tiran to Israeli ships and on 30 May entered into a defence treaty with Jordan. On 5 June, Israel attacked Egypt, Syria, Jordan and Iraq simultaneously and, by the end of the war, six days later, had gained control of the entire Sinai Peninsula and the Gaza Strip and, more importantly, East Jerusalem, the Golan Heights and the West Bank. The area of Israel had therefore been increased from 20,000 km^2 to 70,000 km^2, while the number of Palestinians under Jewish rule had grown from 300,000 to 1.2 million. At this time, the Jewish population was approximately 2.5 million. Despite United Nations Security Council Resolution 242, which demanded withdrawal, Israel retained the territory, thereby intensifying the Arab–Jewish conflict.

In 1970, King Hussein of Jordan, feeling his security threatened by the PLO, ejected the Organization, which moved its centre of activities to Lebanon. By 1973, frustrated by the loss of territory, Egypt and Syria attacked Israel in the Yom Kippur War of 6 October. After initial losses, the Israeli forces again prevailed and, following the cease-fire, a new United Nations Emergency Force was deployed on the Suez Front. With the effective collapse of government in Lebanon during the latter part of the 1970s, the PLO was able to launch a series of attacks on Israel, which retaliated in 1978. Subsequently, a United Nations force was installed in southern Lebanon.

At the same time, President Sadat of Egypt was negotiating with Prime Minister Begin of Israel and on 26 March 1979, a peace treaty was signed between their respective countries, the Camp David Accords. As a result, Israeli right of passage through the Suez Canal was granted and the Gulf of Aqaba and the Strait of Tiran were recognized as international waterways. In response, Israel agreed to vacate the Sinai Peninsula and to begin negotiations towards 'full autonomy' for the Palestinians in the Gaza Strip and the West Bank.

By 1982, Israel had withdrawn from Sinai but, in the same year, invaded Lebanon for the second time in an attempt to eliminate the PLO. Later in 1982, a number of peace initiatives were proposed and in 1983 the Palestine National Council accepted the idea of a confederal relationship between Jordan and Palestine.

Conflicting aims within the PLO led to factional disputes and this, together with mounting Palestinian frustration, resulted, in January 1988, in the *intifada*, a civil uprising which continues to the present. Meanwhile, the PLO acknowledged Israel's right to exist, but refused to halt the *intifada* until an independent Palestinian state had been established.

As a result, the United States began negotiations with the PLO, although Israel refused to follow suit. There ensued an 'on–off' peace process, which was eventually interrupted by the Gulf Conflict of January 1991. During this, Saddam Hussein, leader of Iraq, sought to link his country's occupation of Kuwait with Israel's continuing retention of the Occupied Territories despite United Nations Resolutions 242 and 338. In spite of the complete defeat of the Iraqi forces, the concept of 'linkage' prevailed and, at the end of Operation Desert Storm the United States began strenuous efforts to promote a new Middle Eastern peace programme to settle once and for all the Arab–Israeli dispute.

Status

All three areas have remained under Israeli occupation since 1967. As a result of the gross overcrowding and the high level of conflict between the Palestinians and the Israeli military, the Gaza Strip has become the power base for Islamic Fundamentalism among the Palestinians. However, in the first elections to be held in 25 years, those for the Chamber of Commerce, conducted on 5 November 1991, Fatah-affiliated candidates won 13 of the 16 seats. This must be a sign of support for the continuing peace process.

Economically, problems abound and, in particular, the lack of services and jobs. Furthermore, the excessive use of water by Israeli settlements has put the coastal aquifer under severe pressure, resulting in saline incursions. Geopolitically, the Gaza Strip has little significance and its problems are so severe that it appears to be wanted, in the long term, by neither Israel nor Egypt. It is an area of continuing conflict.

Both in the peace talks and in United Nations Security Council Resolutions 242, 338 and 425, Israeli withdrawal from the Golan Heights is linked to withdrawal from the West Bank and south Lebanon. Indeed, ahead of the Madrid Peace Conference, Syria stated it would not tolerate any deviation from the Resolutions and on 31 October 1991 President Assad asserted 'partial solutions lead to comprehensive wars'. After the Conference, on 12 November, the Knesset passed a resolution, describing Golan as 'an inseparable part of the State of Israel'. This statement was ridiculed by Syria as mere provocation.

The importance of the Golan Heights to Israel is both strategic and economic. Prior to 1967, from such an elevation, Syria was able to mount a constant barrage of small-scale attacks on local Israeli agricultural settlements. This activity has, of course, been eliminated by occupation, which has also allowed the Israelis to command the tributaries of the Jordan and the politically sensitive Yarmuk triangle. They have also established new settlements such as the recently announced Qela, and there appears to be no halt to this process. They have also constructed reservoirs, which are of local rather than national significance. At present, there are some 10,000 settlers in the Golan Heights area and plans were revealed in October 1991 for an increase of some 50 per cent. Given the passionate feelings of Syria, this area is likely to remain a long-term flashpoint.

However, the most contentious area remains the West Bank, together with East Jerusalem which was annexed by the Israelis. The West Bank is seen as a key element of Eretz Israel (Land of Israel), since its constituent parts, Judea and Samaria were, according to the Old Testament, given to the Jews by God. The West Bank is therefore central to any concept of the Holy Land. One result has been a massive settlement policy and, since 1977, it is estimated that some 100,000 Jews have been settled in the West Bank and 120,000 in East Jerusalem. As there has also been much infilling and suburb construction, it is difficult to estimate exact numbers. The original Green Line has been virtually obliterated by settlements built to straddle it and leaked reports indicate that a further 8,000 homes were proposed in the West Bank for immigrants. However the new Rabin government seems unlikely to implement this proposal.

The most compelling economic issue concerns water and, by their occupation, the Israelis gained access to the eastern aquifer and total control over the northern and western aquifers. Since the coastal aquifer, the only one within the original boundaries of Israel, is largely replenished from the West Bank, the importance of the area to Israel for groundwater is obvious. Taking the most conservative estimate, Israel relies for 60 per cent of its needs on groundwater, and therefore the idea of trading 'land for peace' would further endanger the already precarious supplies.

The major result of competition for water supplies and land as an outcome of the new settlements has been continuing conflict between the Palestinians and the Israelis, culminating in the *intifada*. There have been numerous instances of human rights abuses and it is difficult to see any long-term peaceful solution, short of self-government for the Palestinians. Many of the settlers appear to see the answer in the expulsion of the Palestinians. Clearly, between these two extremes there is very little middle ground. In the case of East Jerusalem, Israel has refused to countenance any negotiation, while, for the Palestinians, the city has a spiritual and symbolical significance which elevates it above any negotiation.

Since 1948 and particularly 1967, there has been almost continuous conflict in the West Bank and, given the entrenched positions on both sides, a negotiated solution looks as remote as ever. This is despite the election of a Labour government in Israel in early summer 1992 which may introduce modifications in policy towards the Palestinians.

References

Anderson, E.W. (1988), 'Water: The Next Strategic Resource' in J.R. Starr and D.S. Stoll, *The Politics of Scarcity*, Westview, Boulder, pp. 1–21.

Anderson, E.W. (1988), 'Water Resources and Boundaries in the Middle East' in G.H. Blake and R.N. Schofield (eds), *Boundaries and State Territory in the Middle East and North Africa*, MENAS, Wisbech, pp. 85–98.

Anderson, E.W. (1988), 'The Vulnerability of Arab Water Resources', *Arab Affairs*, 1, No. 7, pp. 73–81.

Anderson, E.W. (1991), 'The Violence of Thirst', *Geographical Magazine*, **LXIII**, No. V, pp. 31–4.

Anderson, E.W. (1991), 'The Middle East and Hydropolitics', *World Energy Council Journal*, December, pp. 35–8.

Anderson, E.W. and Agnew, C.T. (1992), *Water Resources in the Arid Realm*, Routledge, London.

Blake, G.H., Dewdney, J. and Mitchell, J. (1987), *The Cambridge Atlas of the Middle East and North Africa*, Cambridge University Press, Cambridge.

Blake, G.H. and Drysdale, A. (1985), *The Middle East and North Africa: A Political Geography*, Oxford University Press, Oxford.

Boyd, A. (1991), *An Atlas of World Affairs*, Routledge, London.

Day, A.J. (ed.) (1984), *Border and Territorial Disputes*, Longman, London.

Downing, D. (1980), *An Atlas of Territorial and Border Disputes*, New English Library, London.

Munro, D. and Day, A.J. (1990), *A World Record of Major Conflict Areas*, Arnold, London.

26 Guantanamo

Description

Guantanamo is a United States naval-air base which was retained by the United States on mainland Cuba after the Communist revolution led by Fidel Castro, following the Civil War (1956–9) produced an anti-American regime in Cuba. It is therefore particularly interesting as an American enclave in one of the last Marxist–Leninist states in the world. As such, it has proved a long-running source of dispute between the United States and Cuba. Originally ceded to the United States in 1901 following the Spanish–American War, on a lease running until 1999, since independence (1902) and particularly since the revolution, Cuba has consistently demanded Guantanamo's return.

History and importance

Cuba is the nearest of all the West Indian islands to the mainland of the United States, being separated from it by the Florida Strait, a channel some 160 km wide. It has always therefore been a special concern of the United States, which helped to free it from Spanish rule in 1898 during the Spanish–American War. Although, educationally, the most advanced country in Latin America, it has had to endure virtually uninterrupted bad government from independence until the time of Castro.

In 1952, President Batista made himself permanent dictator and instigated a reign of terror. After an initial abortive attempt in 1953, Fidel Castro began the revolution in December 1956. After various setbacks, he was successful on 1 January 1959 when the corrupt regime of Batista finally collapsed. Within a year, the new government began a massive programme of reform, a Communist regime was established and Cuba entered an alliance with the Soviet Union.

Such events horrified the United States, which sponsored the disastrous Bay of Pigs invasion by Cuban exiles in 1961 and then, in perhaps the most serious event of the Cold War, outfaced the Soviet Union in 1962 in the Cuban Missile Crisis. During that period, from 22 October to 20 November 1962, the Guantanamo base played an important surveillance role. Guantanamo's continuing importance derives from its position on an island strategically placed at the entry to the Gulf of Mexico.

Status

American–Cuban relations remain poor and the situation is exacerbated by the large number of anti-communist Cuban refugees now living in Florida. With regard to Guantanamo, the situation remains unchanged. Cuba vigorously prosecutes its claims, while the United States continues to maintain naval and air facilities. Until there is a radical change in relations between the two governments, it is very unlikely there will be any movement on the issue, although the termination of the lease in 1999 and the continuing isolation of Cuba and economic problems caused by the collapse of the Soviet Union (until 1989, Cuba's largest aid donor) may well increase pressure for a settlement.

A regime change in Cuba may now be more likely. Furthermore, the base now has a further problem as it has become a refugee camp for boat people from Haiti, fleeing after the repressive government which followed the coup which ousted President Aristide. American coastguard vessels have rescued some 6,400 boat people since the coup, but the United States had to halt its forced repatriation programme to Haiti in response to growing international and domestic criticism, and there are now some 5,300 Haitians held at Guantanamo base (10 December 1991) more or less indefinitely. The long-term future of these unfortunate refugees is uncertain and their numbers are being added to continually.

Given the uncertainties of Cuban stability, the refugee problem and post-Cold War geostrategic thinking, as well as the imminent termination of the lease, the incoming United States administration will have to consider seriously the position of Guantanamo. Its geostrategic importance to American influence in the Caribbean is obvious, but this has to be balanced against the high cost in political and financial terms of its maintenance.

References

Daily Telegraph (1991) 10 December.
Downing, D. (1980), *An Atlas of Territorial and Border Disputes*, New English Library, London.

27 Guyana

Description

The Cooperative Republic of Guyana, formerly British Guiana, became independent in 1966. It has an area of 215,000 km^2 and a population of 790,000 (1991) which is ethnically divided between Asian Indians (51 per cent), Africans (31 per cent), mixed (12 per cent) and Amerindians (4 per cent). Approximately 90 per cent of the population live in the coastal strip, which is some 500 km long and 16 km wide. Indeed, the coastal plain is the most fertile area, producing sugar cane, the mainstay of the economy, and being responsible for some 70 per cent of the Gross Domestic Product (GDP).

History and importance

Guyana has two long-standing border disputes, with Venezuela in the west and Surinam in the east. The Essequibo dispute (with Venezuela) concerns the territory between the Essequibo River and the Cuyuni–Amakura Rivers which constitute Guyana's western boundary. This is an area of some 140,000 km^2 or approximately two-thirds of the surface area of Guyana. The Essequibo Basin is the heartland around which Guyana developed, and it is also important for resources such as bauxite and possibly petroleum and the Maroni hydro-electric power (HEP) generating site.

In 1648, at the Treaty of Münster, which ended the Spanish–Dutch wars, the Spanish provinces of Essequibo and Berbice were recognized as Dutch. Venezuela (as the successor state to Spain), on the premise that the Essequibo River constituted its 'natural' eastern boundary, complementing the Orinoco River on the west, claimed the area. Furthermore, in disputing the historical claims of the Dutch, it argued that it was the Spanish who first discovered and settled the area.

In the early 19th century, Britain acquired Guyana from the Dutch and later extended its influence west of the Essequibo to borders defined as the 'Schonburgh Line'. The boundaries of what was then known as British Guiana were declared officially in 1866, but the issue was raised again in the 1880s by Venezuela, when Britain made grants of land to residents east of the Schonburgh Line.

In 1895, American President Cleveland stated that the British action violated the Monroe Doctrine, which stipulated no European influence in the Americas. He demanded that the dispute be taken to arbitration.

As a result, in February 1897, at the Paneenfort-Andrade Treaty, Britain and Venezuela agreed to accept as binding the decision of an arbitration panel, consisting of two British and two American delegates, with a Russian president. In 1899, the panel announced its decision, awarding the majority of the territory to Britain, but to Venezuela some 13,000 km^2, including the mouth of the Orinoco River and much of the Cuyuni Basin. Reluctantly, Venezuela accepted the decision at the Treaty of Washington 1899. In 1905, the border was officially demarcated.

The situation remained stable until 1951 when Venezuela questioned the validity of the 1899 decision on the basis of the memoirs, published posthumously in 1949, of an American lawyer who had been a junior on the case. He suggested that a secret deal had been made between Britain and Russia, but this was denied by Britain.

However, in 1962 Venezuela unilaterally declared the 1899 decision to be null and void and on 17 February 1966 the Geneva Agreement was reached between Britain and Venezuela to establish a mixed commission to seek a settlement. Later that year, British Guiana became independent as Guyana; the new nation joined the United Nations, but not the Organization of American States (OAS) because of the latter organization's ambivalent attitude to the dispute. Incidents continually hampered the work of the commission, but in June 1970 it was agreed, through the Port of Spain Protocol, that there should be a 12-year cooling-off

period. During that period, despite an exchange of presidential visits, no progress was made and the Protocol was allowed to lapse. In 1982, Venezuela adopted a strongly pro-Argentinian stance during the Falklands War and there was a great deal of anti-British feeling.

On Guyana's eastern border with Surinam there is the New River Triangle, a triangle of disputed land (approximately 14,500 km^2) between the Courantyne-Cutari Rivers, the New River and Guyana's border with Brazil. Both the Cutari and the New Rivers are tributaries of the Courantyne and the dispute concerns which of them should be the boundary. The area of land at stake is approximately 14,500 km^2. In the 1840s, the border was surveyed by Schonburgh and the Courantyne-Cutari defined as the main course and therefore agreed as the boundary. However, in 1871 the explorer Barrington Brown discovered the New River, which he believed was a larger tributary of the Courantyne than the Cutari. In fact, the Cutari Basin is bigger than that of the New River.

In 1899, the Dutch, who controlled Surinam, claimed the New River as the source of the Courantyne and therefore, their territory's boundary with Guyana. On 4 August 1930, the Dutch offered to settle the problem by accepting a border along the left bank of the Courantyne-Cutari to its source and Britain agreed. However, the final draft of this agreement was not signed, owing to the onset of World War II.

After World War II, the Dutch position hardened and the Courantyne-Cutari line was discarded in favour of the New River. In December 1967, Guyana ejected a Surinam HEP survey team from the triangle and as a consequence Surinam, in turn, threatened to expel all Guyanans. After this, various incidents occurred and in November 1975 when Surinam became independent, it maintained its claim.

Status

Although neither boundary is significant in global terms, the fact that the disputes remain unresolved has resulted in long-term instability in the region. In the case of Guyana's western boundary, there is no pressure with regard to resources on Venezuela, but the outcome of the dispute is fundamental to the survival of Guyana. The eastern boundary is far less critical, but there remains the problem of the maritime boundary at the mouth of the Courantyne.

References

Boyd, A. (1991), *An Atlas of World Affairs*, Routledge, London.
Child, J. (1985), *Geopolitics and Conflict in South America*, Praeger/Hoover Institution Press, Stanford.
Day, A.J. (ed.) (1984), *Border and Territorial Disputes*, Longman, London.
Downing, D. (1980), *An Atlas of Territorial and Border Disputes*, New English Library, London.
The Economist Atlas (1989), Economist Books, Hutchinson, London.
New Scientist (1990) 31 March.

28 The Hatay

Description

The Hatay, also known as the Sanjak of Alexandretta, is the southernmost part of Turkey and its only territory within the Levant. The Hatay appendage measures some 120 km from north to south and about 90 km at its widest, from east to west. It has a population of approximately 750,000 (1975). It comprises essentially a coastal range, the Gavur Daglari, the lowlands around Lake Amik, the massif of the Jebel Akra and, most importantly, the lower course of the Orontes River.

History and importance

Under the terms of the Sykes–Picot Agreement of 16 May 1916, Syria, including the Hatay, was given to France. At the subsequent San Remo Conference in April 1920, the Allies agreed to the French mandate over Syria, including the Hatay. Under the terms of the mandate, France was bound to see that 'no part of Syria and Lebanon is ceded or leased or in any way placed under the control of a foreign power'. On 24 July 1923, the Treaty of Lausanne confirmed this arrangement, together with French control and further proclaimed that Turkey had relinquished all territory left out of her frontiers.

On 9 September 1936, France and Syria concluded a treaty, providing for Syria's independence and territorial integrity. However, France needed Turkish support in the period leading up to World War II and later in that year, a Turkish press campaign began. The centrepiece of this was an allegation that out of the 300,000 population of the Hatay 240,000 were Turkish (together with 25,000 Armenians, 20,000 Sunni Arabs and 15,000 others). French statistics at the time showed a population of 220,000, comprising 85,000 Turkish, 62,000 Alawis, 25,000 Armenians, 23,000 Sunni Arabs, 18,000 Greek Orthodox and other Christians and almost 5,000 Kurds.

One year later, in attempting to appease the Turks, France proposed to give Alexandretta separate administration and on 29 May, a treaty was promulgated, under which only the economic and foreign affairs of the region were to be under Syrian jurisdiction. Later that year, an International Commission of the League of Nations investigated the numbers within the various national groups and reported that Turks accounted for less than 50 per cent of the population. In the November elections, in fact, the Turks were in the minority, but Turkey rejected the Commission's findings.

However, following the exodus of non-Turks, particularly Armenians, and with the added advantage of the French presence, the Turks gained a majority in the elections of 1938. On 4 July, France agreed to the establishment of an 'autonomous republic of Hatay' and permitted French troops to enter the Sanjak of Alexandretta. Elections two months later resulted in the Turks gaining 22 of the 40 seats available. Nevertheless, the presence of Turkish troops undoubtedly encouraged the exodus of non-Turks and helped the pressures for union with Turkey which was immediately claimed by the new parliament. Later in the year, after a plebiscite, the parliament of Hatay proclaimed the incorporation of the district into the Republic of Turkey. Finally, on 23 June 1939, France recognized the transfer of the territory in exchange for a non-aggression pact that ensured Turkish neutrality during World War II.

In economic terms, the Hatay is important because its loss reduced the coastline of Syria by almost a half and also denied it the important areas for irrigation in the lower course of the Orontes. Even more important has been the question of national pride.

Status

The Hatay is now wholly Turkish and fully integrated into the Turkish state. Despite the fact that there is little hope of its ever being returned, or even of its gaining some degree of autonomy, Syria has never accepted the detachment of the Hatay. It remains an important factor colouring Turkey–Syria relations.

References

Arnakis, G.G. and Vucinich, W.S. (1972), *The Near East in Modern Times*, Volume 2, Jenkins, Austin and New York.

Aroian, L.A. and Mitchell, R.P. (1984), *The Modern Middle East and North Africa*, Macmillan, London and New York.

29 The Hawar Islands

Description

Hawar or Huwar, much the largest of a group of 16 barren islands and reefs in the Persian/Arabian Gulf, lies under 2 km from the west coast of Qatar. It is located adjacent to the onshore Qatari oilfield and it is said that it may be reached by foot at low tide. Apart from an occasional token Bahraini military presence, the islands are uninhabited.

History and importance

The dispute over the ownership of the islands dates back to a disagreement over oil concessions in the 1930s between the Bahrain Petroleum Company and Petroleum Concessions Ltd in Qatar. When the ruler of Bahrain set up a small military post on Hawar Island in 1936, the ruler of Qatar complained to the British political agent in Bahrain. Both rulers thereafter presented their claims to the British political resident in the Persian Gulf who subsequently awarded the islands to Bahrain in 1939.

However, Qatar pursued its claim, but the talks over the Bahrain/Qatar offshore boundaries became deadlocked in 1967. In 1976, the issue resurfaced when the foreign ministers of both countries restated their claims and then in March 1978 Qatar detained some Bahraini fishermen, following Bahraini military manoeuvres near the islands. Indeed, the dispute became sufficiently heated that in March 1982 Qatar objected when Bahrain named its new warship *Hawar*.

In February 1981, together with Kuwait, Oman, Saudi Arabia and the United Arab Emirates, Qatar and Bahrain became members of the Gulf Cooperation Council (GCC), one of the functions of which was to settle disputes among member states. Membership exercised a calming influence, and from 1982 the two countries agreed to freeze their differences and accept Saudi Arabian mediation. However, the dispute erupted again in 1986 and, despite Saudi diplomacy, climaxed in the seizure by Qatari forces of 29 foreign workers building a Bahraini coastguard station on one of the islands. Again, in July 1991, tensions rose creating, it was reported, an obstacle to further GCC meetings. As a result, Oman, which supports the claim of Qatar, insisted on international arbitration.

The islands, and especially Hawar itself, are particularly important as a result of resource potential in the area. There are proved reserves of 150 million cubic feet of gas in the North Dome gasfield, some 15 km from the disputed coral reef off Fasht-al-Dibal. Furthermore, in the immediately adjacent continental shelf there is a potentially rich oilfield. As both Bahrain and Qatar are running short of oil reserves, the outcome of the dispute would make a major difference to their economies. Bahrain has very little oil left and is increasingly reliant on diversifying its economy. Qatar is an important gas producer but supplies only modest amounts of oil.

Status

A communiqué of 8 July 1991 from the ICJ reported that Qatar had filed a case against Bahrain over the Hawar Islands. It is known that both sides have teams working on the issues involved, but, if previous procedures are followed, a decision is unlikely in the near future.

References

Blake, G.H. and Drysdale, A. (1985), *The Middle East and North Africa: A Political Geography*, Oxford University Press, Oxford.

Boundary Bulletin (1992) No. 3, International Boundaries Research Unit, Durham University, January.

Day, A.J. (ed.) (1984), *Border and Territorial Disputes*, Longman, London.

Peterson, J.E. (1985), 'The Islands of Arabia: Their Recent History and Strategic Importance' in *Arabian Studies VII*, 3, pp. 23–35.

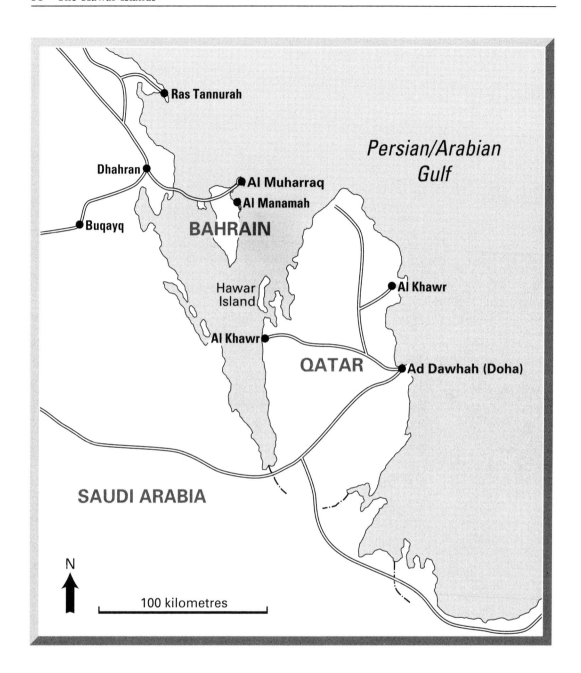

Prescott, J.R.V. (1985), *The Maritime Political Boundaries of the World*, Methuen, London.

Swearingen, W.D. (1981), 'Sources of Conflict over Oil in the Persian/Arabian Gulf', *The Middle East Journal*, **35**, No. 3, Summer, Middle East Institute, Washington DC, pp. 314–30.

Railway to Guangzhou (Canton)

CHINA

GUANGDONG PROVINCE

Shenzhen

Mirs Bay

Deep Bay

NEW TERRITORIES

Tsuen Wan

Kowloon

Lantau Island

Victoria

HONG KONG

South China Sea

Special Chinese economic zone
Built up area

30 Hong Kong

Description

Hong Kong, a British Crown Colony, comprises three elements:

(a) Victoria or Hong Kong Island, together with a number of nearby islands, the largest being Lantau;
(b) Kowloon, the urbanized tip of the Chinese mainland peninsula, facing the city of Victoria across the harbour; and
(c) the New Territories, the remainder of the peninsula from Kowloon to the border with China.

Hong Kong Island itself lies just over 30 km east of the mouth of the Pearl River and 146 km southeast of Guangzhou (Canton). It covers an area of 75.9 km^2 and is separated from Kowloon (10.35 km^2) by one of the finest natural deep water harbours in the world. The New Territories, including areas reclaimed, cover some 1,056 km^2. The population of the colony is 5.9 million, of whom 98 per cent are Chinese and the GDP is HK $736.3 billion (US $94.4 billion).

Hong Kong is largely dependent for its food and almost totally dependent for its water on mainland China. Militarily, Hong Kong is entirely at China's mercy. Economically, it has become integrated with China's hinterland and provides China with some 40 per cent of all its foreign-exchange earnings. This encouraged the establishment of the adjacent Special Economic Zone in Guangdang province. Hong Kong is a hub of sea and air links and is linked by rail to Guangzhou.

History and importance

During the 1830s, British traders pressed the government for military and naval assistance to open the Chinese market to the opium trade. This led to the first Opium War (1839–42), which resulted in the retreat of the British community from Canton to the Portuguese outpost of Macao and, in August 1839, to Hong Kong Island. In 1841, there was an attempt by British military commanders to confirm the British occupation of Hong Kong, but this was rejected by both governments. However, the potential value of the island so impressed the British Government Representative, Sir Henry Pottinger, that he allowed the colony to become a *fait accompli* against the wishes of the foreign secretary and prior to the Treaty of Nanking (1842), which ended the war.

Having annexed Hong Kong Island from the Manchus, during the 1840s, British and American merchants began construction on the mainland at Kowloon. Various disputes followed and eventually further hostilities between Britain and China, the latter also being under threat as a result of the Taiping Rebellion. The outcome was that in March 1860, the governor of Hong Kong was able to extract the lease of the Kowloon Peninsula. In October 1861, Britain officially annexed the Kowloon Peninsula, northwards to the line of 'Boundary Street' and Stonecutters Harbour.

Thirty years later, in 1893, Britain felt that Hong Kong was threatened as a result of the Franco–Russian Alliance and therefore pressed for possession of an area inland of Kowloon. In 1895, having been defeated by Japan, China was weak and Britain was able to acquire a 99-year lease of the New Territories in a negotiation which was concluded on 1 July 1898. The lease expires in 1997. In the same agreement, Britain unilaterally cancelled the clause, formally agreed, that China should retain, as an enclave, the old walled city of Kowloon. The New Territories provided additional security, more space to settle the burgeoning population from Victoria and Kowloon and an area to grow large quantities of food.

From 1899, when Britain officially took possession of the New Territories, until 1949, communications across the border between Hong Kong and China were almost unimpeded. However, with the establishment of the People's Republic of China, severe restrictions ensued. These were only effectively loosened in August 1980 when Shenzhen, an area of 327.5 km², just north of the Colony, was selected as a Special Economic Zone.

By 1982, it was clear that China had adopted a policy of 'one country, but two systems' and this clearly improved the prospects for the future of Hong Kong. In September 1984, a Sino–British agreement allowed for Chinese sovereignty to be re-established over the whole of Hong Kong in 1997. The former Colony would be merged with Shenzhen after that date. Under the agreement, Beijing promised to leave the existing economic and social systems essentially unchanged for a period of at least 50 years, as well as to permit a degree of self-government. However, there is no way that these assurances can be guaranteed and there has been wide-scale emigration from Hong Kong, particularly by the wealthy and better educated.

Status

The major problem is that fears of what might happen after 1997 have led to a capital outflow and brain-drain, thereby undermining the economy, which is stagnating. In early 1991, inflation reached 10 per cent. Furthermore, China has shown an increasing desire to influence events in Hong Kong, such as becoming involved in the development of the new airport, and, although confrontation has been avoided, confidence in the future has been severely undermined. Friction has also been generated as a result of moves towards full democracy and a representative government in the Colony. Lack of democracy angers Hong Kong, moves towards it infuriate China.

As 1997 nears, and, in the light of economic problems in China, there must be a fear of large-scale illegal immigration into Hong Kong from China. It would appear to be in China's interests to maintain Hong Kong as the third largest financial centre and the fourth largest container port in the world, but, current ideology is such that the acquisition of the territory seems to be more important than the maintenance of any economic benefits that might naturally accrue from it. While there may be minor conflicts before 1997, Hong Kong, unsupported by any outside power, will have to acquiesce in the take-over and is extremely unlikely to become a geopolitical flashpoint, though in global terms it may spark off instability in China itself.

References

Boundary Bulletin, No. 1, (1991), International Boundaries Research Unit, Durham University.
Boundary Bulletin, No. 3, (1992), International Boundaries Research Unit, Durham University.
Boyd, A. (1991), *An Atlas of World Affairs*, Routledge, London.
The Economist, The World in 1992, Economist Publications, London.

31 The Strait of Hormuz

Description

Since the Iranian Revolution in 1979, the Strait of Hormuz has become by far the best known choke point in the world. This high profile was reinforced by the Soviet invasion of Afghanistan in 1979, following which there were widespread fears that the Strait might be closed to shipping. However, the main channels are 70-90 m deep and therefore the physical act of attempting closure would cause problems, while Iran, the country most vociferous in its threats, is the one in the region which could least afford closure.

The Strait of Hormuz comprises a curved channel, connecting the water of the Persian/Arabian Gulf with those of the Gulf of Oman and the Indian Ocean. It therefore offers the only maritime access to the Gulf and is crucial to such states as Kuwait, Bahrain and Qatar, which have no other outlet for their oil, as well as to Iraq, Iran and Saudi Arabia. It is approximately 100 nml long and 39 km wide at its narrowest point between Larak Island (Iran) and the Quoin Islands (Oman). Thus, as a result of its ownership of the Musandam Peninsula, a detached part of its territory, Oman has a vital role in ensuring free passage through the Strait. As both Oman and Iran claim 12 nml territorial waters into the Strait, their respective claims are delimited by a 28-km-long median line.

History and importance

The Persian/Arabian Gulf region accounts for 65.6 per cent of world oil reserves and 26.5 per cent of production (1990). While Europe and the United States receive an increasing proportion of their oil via pipelines, allowing Hormuz to be bypassed, all oil from the Gulf for Japan travels by tanker through the Strait. For the major industrialized nations, the choke point of Hormuz is therefore vital. Furthermore, it is so crucial for Iran that the Shah referred to the Strait as Iran's 'jugular vein'.

Between 1979 and 1988, the Strait acquired an even higher profile as a result of threats, mainly Iranian, to close it. Between 70 and 80 ships, including the largest super-tankers, transit the Strait daily as part of the vital supply system for the industrialized world. In 1990, the Gulf region was a routeway for 25 per cent of United States' imports, 41 per cent of West European imports and 68 per cent of Japanese imports. Furthermore, in the context of the increasing dominance of the Pacific Rim in world trade, the region was responsible for 83 per cent of Asian (excluding China and Japan) imports. With declining petroleum production elsewhere, and with limited spare capacity in countries such as Mexico, Venezuela and Nigeria, the role of the Gulf as an oil supplier can only increase. Indeed, given the apparent weak state of the oil industry in the former Soviet Union, greater calls are likely to be made upon this region earlier than had originally been supposed.

Among the key suppliers, the Strait is of particular importance to Iran, which has no alternative export route. Countries such as Iraq, Kuwait and Saudi Arabia can bypass the 'Hormuz factor' by pipelines, notably Petroline, across the Arabian Peninsula to the Red Sea, and the two Dortyol pipelines from Iraq to the Mediterranean via Turkey. Various plans have been advanced to construct an Iranian pipeline along the northern side of the Persian/Arabian Gulf to Hormuz or beyond, but, as yet, nothing has been built.

Threats to Hormuz surfaced regularly after the Iranian revolution and the Soviet invasion of Afghanistan (both in 1979) and during the Iran-Iraq war (1980-8). Indeed, Iranian threats to close the Strait in response to Iraqi attacks on third-party shipping loading at Iranian oil installations, were defused only by joint American-European action.

During the Iran-Iraq war, over 200 ships were damaged. In 1982, Iran threatened that if the Kharg Island terminal were destroyed, Hormuz would be

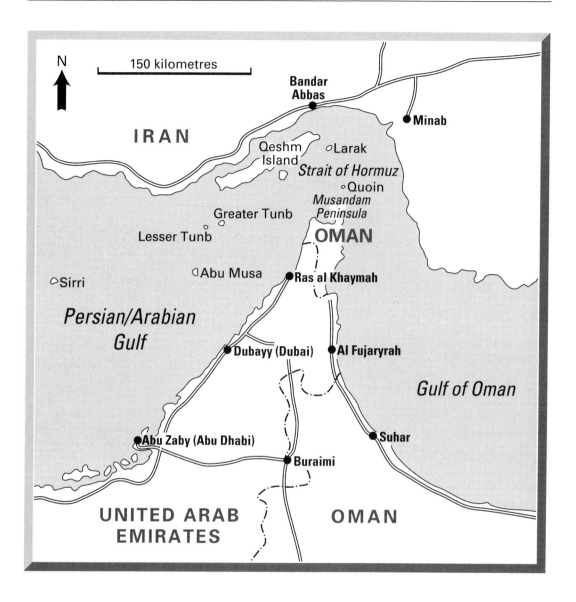

closed by a 'wall of fire' provided by 130 mm guns on islands in the Strait, air-to-air missiles or the sinking of a large oil tanker. In 1984, Iraq made a number of air attacks on Kharg Island and Iran retaliated against ships using Kuwaiti and Saudi ports. This phase of the war, the 'tanker war', became steadily more violent. In 1984, 51 ships were attacked, and by 1987 the figure had risen to 178. Some ships were damaged by mines, some by Iranian missiles and some by fast patrol boats, manned by Iranian Revolutionary Guards. As a result, in 1987 the United States reflagged some Kuwaiti tankers and escorted them through the Gulf in convoy, and early casualty being the tanker *Bridgetown* struck by a mine on 24 July 1987. At the same time, American but mainly British and French minesweepers were sent to the Gulf, joining a fleet of some 75 foreign war-ships.

Iran renewed its threats against Hormuz, having installed Silkworm land-based missiles, a copy of the Soviet 'Styx', near Hormuz. These missiles, with a 50-mile range and 1,000-pound warhead obviously posed a severe threat. Again, the maritime powers, including the Soviet Union, deterred Iran by their angry reaction and the missiles were not used. However, the potential disruption which could be caused by one state with limited maritime forces had been amply demonstrated. One other significant result was the support for Iraq which resulted in United Nations' pressure for a cease-fire being directed mainly at Iran. Even when the *USS Stark* was attacked on 17 May 1987, by a missile from an Iraqi fighter, and 37 men died, support for Iraq did not waver. This position, of course, changed abruptly in August 1990.

Status

Despite its catastrophic losses during its war with Iraq, Iran regained a great deal of its international prestige during Operation Desert Storm. It remains the dominant regional power in the area of Hormuz. However, following Operation Desert Storm, there is likely to be a strong American or United Nations' presence in the region for at least the medium term. Therefore, it is unlikely that meaningful threats against Hormuz can be renewed. Furthermore, in May 1990, Iran and Oman signed an agreement to set up a joint committee to investigate the exploitation of a joint oilfield in the Strait of Hormuz itself.

References

Alexander, L.M. (1988), 'Choke Points of the World Ocean: A Geographic and Military Assessment' in E.M. Borgese, N. Ginsburg and J.R. Morgan, *Ocean Yearbook* 7, University of Chicago Press, Chicago, pp. 340-55.

Anderson, E.W. (1985), 'Dire Straits', *Defense and Diplomacy*, 3, No. 9, pp. 16-20.

Blake, G.H., Dewdney, J. and Mitchell, J. (1987), *The Cambridge Atlas of the Middle East and North Africa*, Cambridge University Press, Cambridge.

Blake, G.H. and Drysdale, A. (1985), *The Middle East and North Africa: A Political Geography*, Oxford University Press, Oxford.

Boundary Bulletin, No. 1, (1991), International Boundaries Research Unit, Durham University.

Hiro, D. (1985), *Iran Under the Ayatollahs*, Routledge & Kegan Paul, London.

Mojtahed-Zadeh, P. (1990), 'Iran's role in the Strait of Hormuz 1970-1990'.

Peterson, J.E. (1985), 'The Islands of Arabia: Their Recent History and Strategic Importance' in Arabian Studies, VII, London, pp. 23-35.

Rais, R.B. (1986), *The Indian Ocean and the Superpowers*, Croom Helm, London.

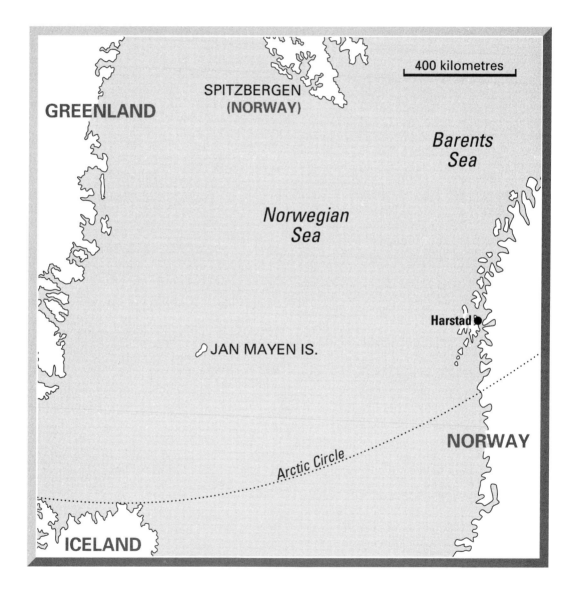

32 Jan Mayen Island

Description

Jan Mayen Island 55 km long, with an area of 380m², is located 500 km northeast of Iceland and is the only island situated centrally in the Norwegian Sea. It comprises an extinct volcano, rising to 2,277 m, covered in glaciers. There are no native inhabitants, but there has been a small transient scientific population, and Norway established a weather station on the island in 1921. Jan Mayen was formally annexed by Norway on 8 May 1929.

History and importance

For most of the post-World War II period, Jan Mayen has been important as a centre for the NATO surveillance of Soviet movements, particularly those involving the Northern Fleet. This may now be a point of historical significance only, although there appear to have been no large-scale reductions in the Northern Fleet, and how Russia might use it must remain conjectural.

Of continuing importance is the location of Jan Mayen in the centre of very rich fishing grounds. On 22 December 1976, Denmark declared a 200 nml fishing zone around Greenland, although, on the east coast, this terminated at latitude 67°N, approximately the location of northern Iceland. On 29 May 1980, Norway declared a 200 nml fishing zone around Jan Mayen. Three days later, Denmark replied with a proclamation extending Greenland's fishing zone north of 67°N.

Indeed, in 1981, Denmark stated that 'where the island of Jan Mayen lies opposite Greenland, the extent of the fishery zone is 200 nml'. If this were the accepted situation, Jan Mayen would be left with no fishing zone. The special circumstances advanced by Denmark in support of its claim included Greenland's dependence on fishing, Jan Mayen's remoteness from Norway and the small size and non-indigenous nature of the population. Against this, it can be said that the fishermen of Greenland had previously showed little interest in the area. Furthermore, Jan Mayen can clearly be defined as an island, rather than a mere rock such as Rockall, under the terms established by the Law of the Sea Convention.

To further complicate the dispute, in 1981 Iceland and Norway established a Conciliation Commission, which concluded that the sea-bed was a 'microcontinent' and therefore could not be regarded as the natural prolongation of either Jan Mayen or Iceland. It therefore recommended joint exploitation in an area three-quarters on the Jan Mayen side of Iceland's 200-mile EEZ and one-quarter within Iceland's claim. In the latter area, Norway would acquire a 25 per cent stake in any venture, while Iceland would receive 25 per cent in the former area.

Status

Given the good relations generally existing between the disputants, armed conflict over Jan Mayen is highly unlikely. However, Denmark's claim to Icelandic waters appears, by any standards, excessive. The rights of small islands located in large areas of open sea are likely to come under increasing scrutiny as the wealth of offshore areas is revealed. If the ICJ considers equity, then there are difficulties in awarding to a small piece of land an area many hundreds or even thousands of times its own size. This is particularly the case when other states in contention are large. Malta might be cited as a parallel case. However, such disputes are likely to remain dormant rather than active flashpoints.

References

Armstrong, T., Rogers, G. and Rowley, G. (1978), *The Circumpolar North*, Methuen, London.
Blake, G.H. and Anderson, E.W. (1982), *International Maritime Boundary Delimitation and Islands: Examples Analogous*

to Libya–Malta, University of Durham (unpublished), September.

Boyd, A. (1991), *An Atlas of World Affairs*, Routledge, London.

Leighton, M.K. (1979), *The Soviet Threat to N.A.T.O.'s Northern Flank*, Agenda Paper No. 10, National Strategy Information Center, Inc., New York.

Prescott, J.R.V. (1985), *The Maritime Political Boundaries of the World*, Methuen, London.

33 Karelia

Description

Karelia is an autonomous republic of Russia which has long functioned as a buffer zone between Finland and Russia, and before 1991, the former Soviet Union. It stretches along the frontier dividing the two countries from the Gulf of Finland to the Kola Peninsula. It includes approximately half of Lake Ladoga and its principal city is Vyborg. It consists of forests, marshes and lakes and is celebrated in his music by the Finnish national composer Jean Sibelius.

History and importance

For some 400 years, Finland itself was a buffer zone between two great powers: Russia and Sweden. In the early 1300s, the Swedes began to colonize lower Finland and eventually, during the 15th and 16th centuries, a loose Swedish–Finnish confederation was formed. In 1617, following a successful campaign, Sweden was able to annex the Karelian districts of Kakisalmi and Inkeri. However, later, the Great Northern War (1700–21) and subsequent disastrous campaigns led to their return to Russia. Eventually, the Russian–Swedish conflict of 1807–09, which followed Napoleon's Truce of Tilsit, ended with the Treaty of Hamina and the complete cession of Finland to Russia.

Nevertheless, 90 years of Russification were unsuccessful and on 6 December 1917 Finland declared itself independent under the constitution of 1772. On 4 January 1918, Russia, followed by France, Germany and Sweden, recognized the independence of Finland. Later, in 1920, in the Treaty of Dorpat, the boundary was renegotiated and delimited and Finland abandoned claims for Russian Karelia in exchange for Petsamo (Pechenga) and a corridor to the Barents Sea. The border was surveyed and demarcated from 1920 onwards and the final protocol signed in 1938.

In 1939, however, the Soviet Union demanded a mutual assistance pact, territorial concessions in Karelia and the north and the lease of the naval base at Hango. The Finns refused and on 30 November 1939 the 'Winter War' broke out. On 12 March 1940, Finland signed an armistice, which led to the loss of eastern Karelia, Salla and Petsamo and the lease of Hango. The territory lost included 10 per cent of the Finnish population, as well as a major port, Vyborg (Viipuri).

During World War II, Soviet attention was directed elsewhere and Finland was able to attain its pre-war boundaries, but on the defeat of Germany, it was forced to sue for peace and to return to its 1940 boundaries. In total, Finland was forced to cede one-eighth of its territory to the Soviet Union and, following a 1948 treaty, was committed to help defend the Soviet Union if it were attacked across Finnish territory. Porkkala, the naval base near Helsinki, which had been occupied in 1944, was handed back in 1956.

The importance of Karelia as a buffer zone has been illustrated by the number of times it, or part of it, has changed hands over the past 600 years. For the Soviet Union, it was particularly important in safeguarding the approaches to Leningrad (St. Petersburg), a fact illustrated by Stalin's willingness to sacrifice approximately 1.5 million Soviet troops in the Finnish campaign.

Status

On 10 August 1990, the Karelian Autonomous Soviet Socialist Republic issued a declaration of sovereignty. As yet, Karelia remains divided by the Finno–Russian border, but, with the vast changes occurring in the former Soviet Union, some measure of reunification must be a possibility.

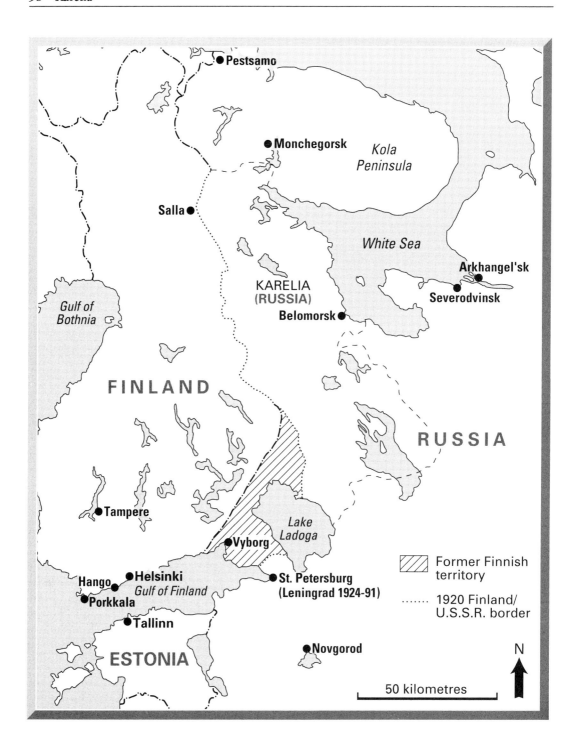

References

Boundary Bulletin, No. 1, (1991), International Boundaries Research Unit, Durham University.

Hamilton, N. (1990), 'The Price of Independence' in *Frontiers*, BBC Books, London, pp. 142–67.

Leighton, M.K. (1979), *The Soviet Threat to N.A.T.O.'s Northern Flank*, Agenda Paper No. 10, National Strategy Information Center, Inc., New York.

United States Department of State (1967), *Finland – U.S.S.R. Boundary*, International Boundary Study No. 74, Office of the Geographer, Bureau of Intelligence and Research, Washington DC, 1 February.

34 Kashmir

Description

Prior to 1947, the state of Jammu and Kashmir had an area of 220,000 km² and a population of just over four million (1941 census). The population comprised 77 per cent Muslims, 20 per cent Hindus, 2 per cent Sikhs, Buddhists and others. The Vale of Kashmir to the south was overwhelmingly Muslim, the Hindu majority was located around Jammu in the southwest and the Buddhists were mainly in evidence in Ladakh, in the north. Since independence from Britain in 1947, there has been an almost continuous state of tension in the region, frequently the focus of the long-term animosity between Pakistan and India.

History and importance

The idea of a permanent separation between the Muslims and the Hindus became policy when it was endorsed by the Muslim League in 1940. In 1947, the predominantly Hindu Congress Party and the Muslim League proved unable to agree on the terms for a draft constitution for an independent India and, as a result, in June that year, the British government declared its intention to grant dominion status to what would be two separate countries, India and Pakistan. To achieve this, the districts with Muslim majorities – British India, Bengal and the Punjab – were to be partitioned; the remaining princely states were to be offered the chance to accede either to India or Pakistan. Independence within the Commonwealth was granted to both countries in August 1947.

Despite the overwhelming Muslim majority in Kashmir, the Maharajah of Jammu and Kashmir could not decide which way to accede and eventually, in October, was forced to join India. Pakistan immediately invaded and Indian forces occupied the eastern portions of Kashmir, including the capital, Srinagar. In January 1949, a cease-fire took place and the Karachi Agreement established a line of control between the two countries. As a result, India maintained its control over the Vale of Kashmir to the south of the cease-fire line, while Pakistan exercised authority over areas to the north and west. Throughout the 1950s, there were regular threats from both sides, but they were able to agree in 1960 to the equitable distribution of the Kashmiri waters under the terms of the Indus Water Treaty.

In 1962, China began raising further border issues. On 3 May, Pakistan and China signed an agreement, demarcating the frontier between the part of Kashmir held by Pakistan and Xinjiang Province. Effectively, Pakistan abandoned its claim to almost 34,000 km² of territory. However, this action was contested by India, and in July fighting broke out between Indian and Chinese forces in the Karakoram mountains.

Between 1964 and 1966, there was further fighting between Pakistan and India, particularly in the mountainous regions to the northeast, but, in January 1966, the Tashkent Declaration, following a cease-fire, allowed troops to be withdrawn from the line of control. During 1971, there were further hostilities between India and Pakistan which accompanied the break-up of Pakistan and the emergence of the Indian-supported Bangladesh in what had been East Pakistan. In 1972, a new line of control between India and Pakistan was delineated by the Simla Agreement. During the 1980s, attention focused on the mountainous areas of Kashmir and in particular, the region of the Siachen Glacier, on which Indian troops had occupied strategic locations. Relations simmered for the remainder of the 1980s and violence finally escalated in early 1990.

The nearly half-century of continuing violence indicates that the conflict is over more than mere territory. The problem is critical for India because the dispute internationalizes India's internal ethnic and religious differences and the survival of the Union itself seems to hang in the balance. To

Pakistan, Kashmir is important in nationalist, military and Islamic Fundamentalist politics. The issue is therefore very deep-seated for both countries.

Status

The current state is one of escalating violence, human rights violations and cross-border conflict. The latest clashes occurred towards the end of August 1991, when almost 30 people were killed. From 3 September, a cease-fire was implemented, but, as on all previous occasions, it will probably only be temporary. Both countries have severe economic problems and both are thought to have nuclear weapons. Therefore, Kashmir is potentially one of the most dangerous flashpoints in the world.

References

Boundary Bulletin, No. 1, (1991), International Boundaries Research Unit, Durham University.

Boundary Bulletin, No. 2, (1991), International Boundaries Research Unit, Durham University.

Boundary Bulletin, No. 3, (1992), International Boundaries Research Unit, Durham University, January.

Boyd, A. (1991), *An Atlas of World Affairs*, Routledge, London.

Clad, J. and Bowring, P. (1990), 'Limits of Tolerance', *Far Eastern Economic Review*, 17 May, pp. 10–12.

Clad, J. (1990), 'Valley of Violence', *Far Eastern Economic Review*, 24 May, pp. 22–3.

Day, A.J. (ed.) (1984), *Border and Territorial Disputes*, Longman, London.

Downing, D. (1980), *An Atlas of Territorial and Border Disputes*, New English Library, London.

The Economist (1991) 13 July.

The Economist (1991) 7 September.

Munro, D. and Day, A.J., (1990), *A World Record of Major Conflict Areas*, Arnold, London.

Osmaston, H. (1990), 'The Kashmir Problem', *Geographical Magazine*, June, pp. 16–19.

35 The Kola Peninsula

Description

The Kola Peninsula was the most important military territory of the Soviet Union and remains of great military potential for Russia. Located at the far northwestern extremity of the Russian land mass, the peninsula measures approximately 600 km by 300 km. Bounded by the Barents Sea to the north and the White Sea to the south and east, it provides the only naval outlet to the west which is not constrained by narrow straits or under the direct surveillance of NATO countries.

Russia's other European ports pose a variety of difficulties. Apart from the problems of icing in the Gulf of Finland, the Baltic Fleet has to transit the Danish straits, while the Black Sea Fleet is constrained by the Turkish straits. The Northern Fleet, based in the neighbourhood of Murmansk, can use ice-free deep-water approaches to move into the Norwegian Sea and North Atlantic. Furthermore, elements of the ballistic-missile submarine fleet can sail northwards, virtually undetected, to take up position under the Barents Sea ice.

The population of the peninsula has tripled since World War II to approximately one million, while that of Murmansk has doubled to 300,000. The region, which is linked to St. Petersburg by railway, accounted for 20 per cent of Soviet fish products (statistics for Russia are not yet available), and it is rich in metals, timber and raw materials for the fertilizer industry. The major industries are military-related, including shipbuilding and ship repairing.

History and importance

As a result of the 'Winter War' between the Soviet Union and Finland, which lasted from 30 November 1939 until the armistice on 17 March 1940, Finland was forced to cede eastern Karelia, comprising the Salla section, just south of the Kola Peninsula and part of the Rybachi Peninsula in the Petsamo region.

At the end of World War II, the whole of the Petsamo (Pechenga) area was ceded to the Soviet Union under the 1944 armistice and confirmed by the 1947 final Treaty of Peace between the two countries. Pechenga, in particular, was important as it is an ice-free port, allowing development as a principal military base. It also is a major fishing port and the site of nickel, copper and uranium mining. Also included, near the tripoint with Norway, was the hydro-electric power (HEP) site at Janiskoski.

The importance of the Kola Peninsula to Soviet and now Russian defence cannot be overstated. The greater part of NATO's maritime effort was directed at nullifying the advantage which the peninsula provided. The peninsula has been described as the most strongly defended military base in the world. The Northern Fleet, which has undergone little change since the demise of the Soviet Union, consists of some 40 ballistic-missile submarines, 140 other submarines, an aircraft carrier and over 80 major surface warships, together with 300 combat aircraft, based at some 40 airfields and a naval infantry brigade. There are also three army divisions stationed on the Kola Peninsula.

Status

Future development depends, of course, crucially upon the military requirements of Russia. It seems reasonable to suppose that there will be an emphasis on the development of an economic infrastructure, industrialization and, probably, further colonization. These possible changes have to be viewed against the background of a potential military rundown and the high levels of pollution, resulting from the present nickel industry.

References

Leighton, M.K. (1979), *The Soviet Threat to N.A.T.O's Northern Flank*, Agenda Paper No. 10, National Strategy Information Center, Inc., New York.

Luton, G. (1986), 'Strategic Issues in the Arctic Region' in E.M. Borgese and N. Ginsburg (eds), *Ocean Yearbook 6*, University of Chicago Press, Chicago, pp. 399-416.

'The Fallout of an Arctic Thaw', *Boundary Bulletin*, No. 1, (1991), International Boundaries Research Unit, Durham University.

United States Department of State (1967), I.B.S., *Finland - U.S.S.R. Boundary*, 1 February.

36 Kosovo

Description

Officially, an autonomous province of the Yugoslav Federation, Kosovo is located between Serbia, Montenegro, Albania and Macedonia. It is a mountainous, economically very backward area, which has received much development assistance but remains poor. It has the highest birth and death rates in the former Yugoslavia and the highest population density, which amounted to 140 per square km (1977), as against 86 for the Federation as a whole. In 1977 just under 1.5 million, was approximately 75 per cent Albanian. As a result of the Albanian birth rate and the Serbian exodus, the proportion is now even more unbalanced. In 1989, there were estimated to be 1.7 million Albanians in Kosovo and under a quarter of a million Serbs.

History and importance

Since the demise of the Yugoslav Federation, Serbia has effectively taken over the government of this Albanian-dominated region, which has, throughout a large part of its history, seen the struggle of Albanian irredentism and Serbian expansionism. However, since at least the 14th century, the history of Kosovo has been intimately associated with all the major events of the region. For example, in 1389, the Turks defeated a coalition of Serbs, Albanians, Bosnians and Wallachians at the Battle of Kosovo.

Nevertheless, the focus must be upon modern times and particularly the period since 1929, when the word 'Yugoslavia' first appeared on the world map. In 1939, Italy annexed Albania and in 1941, the Axis powers (Germany, Bulgaria, Italy and Hungary) invaded Yugoslavia. In November of that year, the Yugoslav Communists established the resistance movement, the Partisans, and helped set up a Communist Party in Albania. At the same time, the region of Kosovo–Metohija, formerly part of Serbia, was integrated with Albania, under Italian administration.

In 1944, Partisans under the leadership of Marshal Tito, liberated Albania and Yugoslavia, Communist regimes took power in both countries and the Kosovo–Metohija region was returned to Serbia. However, the close relationship between Albania and Yugoslavia was short-lived, terminated by the latter upon the expulsion of Yugoslavia by Stalin from the Cominform in 1948. There followed a number of border incidents, but in 1953 relations between the two countries were normalized and the border was demarcated. This was but a temporary lull and several incidents, and rumours of incidents, occurred during the remainder of the 1950s.

In 1963, Kosovo–Metohija was given the status of an autonomous province within Serbia. In 1968, the region changed its name to Kosovo and further unrest took place, with demands that the Republic become autonomous and largely governed by Albanians. During the 1970s, nationalist feelings spread throughout the provinces of Yugoslavia, leading to further clashes between Albanian and Serbian loyalists. However, in 1978, when Chinese support for Albania ended, attempts were made to develop closer economic contact between Albania and Yugoslavia.

In May 1980, President Tito died and the potential for the fragmentation of Yugoslavia increased. It seems probable that for the remainder of the 1980s, the threat of the Soviet Union, alone, maintained the integrity of Yugoslavia. During the period, however, there was continuing unrest in Kosovo and martial law was imposed there in 1987. In 1989, a decision of the Serbian Republican Assembly to extend its control over Kosovo led to further large-scale protests and strikes, but in March of that year the Kosovo provincial assembly endorsed the measures extending Serbian control over the internal affairs of Kosovo and Vojvodina. This action resulted in another crescendo of violence and in July 1990 the Kosovo parliament was suspended when delegates proclaimed independence from Serbia.

The problem of Kosovo is clearly more than a mere territorial dispute. The area has long been considered the cradle of Serbia since, in the 12th century, the Serbian state originated in Kosovo. None the less, the current imbalance in the population amply reinforces the Albanian case.

Status

Following the declaration of independence on 2 July 1990, the Serbian assembly dissolved the Kosovo assembly permanently. The proclamation of independence was denounced as an attempt to create a greater Albania, but the result has been merely to inflame Albanian opinion. While Serbia retains its military superiority within the region, the status quo is likely to be maintained, but a further breakdown in Yugoslavia could lead to the unification of Kosovo and Albania.

References

'Yugoslavia', *Boundary Bulletin*, No. 1, (1991), International Boundaries Research Unit, Durham University, pp. 24–6.

Day, A.J. (ed.) (1984), *Border and Territorial Disputes*, Longman, London.

Griffin, M. and Ward, S. (1989), 'Albanians and Serbs – The Conflict Continues', *Geographical Magazine*, May, pp. 21–4.

Munro, D. and Day, A.J. (1990), *A World Record of Major Conflict Areas*, Arnold, London.

37 Kurdistan

Description

Kurdistan, an area of some 191,600 km², contains a basically homogeneous population of ethnic Kurds and straddles the boundaries of five countries. The total population numbers approximately 20 million and is composed of about 10 million in Turkey, 5 million in Iran, 4 million in Iraq, 500,000 in Syria and 200,000 in the former Soviet Union. The Kurds are not only the fourth most numerous people in the Middle East, but probably the largest nation in the world which has been denied an independent state.

Since it is not a state, the boundaries of Kurdistan represent the approximate limits of Kurdish settlement. The area is strategically located among the mountains at the global crossroads. The Kurds are a mountain people who have had a difficult relationship with the inhabitants of the surrounding plains. As, in Iraq, the nearby lowlands include the major oilfield at Kirkuk, the central government has been particularly diligent in checking any potential Kurdish advance.

For the past 70 years, the Kurds have fought to obtain greater autonomy within their various states, while retaining the ultimate goal of an independent Kurdistan. Within the region, conflict between neighbouring countries such as Iran and Iraq has involved the Kurds, usually as pawns. In each country, the problems for the Kurds differ and, as a people, they have been unable to achieve a lasting unity. Given this and the lack of support by any major power, there appears to be little chance of achieving statehood.

History and importance

The Kurds are descendants of an Indo–European tribe, which settled west of the Caspian Sea some 4,000 years ago. Despite the cultural and political influences of their powerful neighbours, the Persians, the Arabs and the Turks, the Kurds have developed a distinctive identity. The majority are Sunni Muslims and, although they have their own language, Kurdish, the different regional dialect groups cannot communicate freely with each other.

Squeezed between the Ottoman and the Persian Empires, the relationship between the Kurds and their neighbours has been one of almost continuous confrontation, particularly since the early-19th century. In 1826, 1834 and 1853, the Kurds rose against their Turkish rulers and, after World War I and the disintegration of the Ottoman Empire, it did appear that self-determination might be possible. The Treaty of Sèvres (1920) not only established the British and French mandates in the Middle East, but also envisaged local autonomy in Kurdish areas and the possible creation of a Kurdish state. However, two years later, with the abolition of the Turkish Sultanate and the rise of nationalism under Kemal, the Treaty was revised. In 1923, it was effectively replaced by the Treaty of Lausanne in which there was no mention of an independent Kurdistan. None the less, the Treaty of Sèvres has remained a crucial reference point for Kurds in their appeals for global support.

During the 1920s and 1930s, there were further Kurdish revolts against their host nations, particularly Turkey and, after 1932, the newly independent Iraq. World War II brought yet further changes and, following the British and Soviet occupation of Iran, in December 1945, the Kurdish Republic of Mahabad was created. However, by May of the following year, the Soviet troops had withdrawn and the Iranian army had crushed the new Republic.

In the post-war period, Iraq, Iran and Turkey all pursued essentially anti-Kurdish policies, but their alliance was disrupted by the Iraqi revolution of 1958. President Kassim, the new ruler of Iraq, initially supported Kurdish aspirations, but good relations eventually foundered and, with the help of Soviet arms, the Iraqi army mounted a ferocious campaign against the Kurds in 1961–2. With the overthrow of Kassim in February 1963, there was a

short hiatus before operations against the Kurds were resumed, this time under the new Ba'ath regime. At the same time, since both Iraqi regimes were seen as a threat by the Shah, Iran offered support to the Iraqi Kurds.

The situation was reversed some 20 years later when, following the fall of the Shah in 1979 and the subsequent anti-Kurdish onslaught by Ayatollah Khomeini, Iraq supported Kurdish separatists within Iran. During the same period, there were frequent clashes in Turkey and in 1984 martial law was introduced in the Kurdish areas. By 1985, Kurdish separatists were clashing with forces in Turkey, Iraq and Iran and a number of border agreements, restricting Kurdish movement, were made. The most serious incident occurred in 1988 when the Iraqi government responded to Kurdish advances by bombing the town of Halabja with chemical weapons. An international outcry followed the deaths of some 5,000 people as a result of this action.

This virtually continuous conflict drew the plight of the Kurds to global attention but also cost thousands of lives and led to massive refugee movements. In October 1989, leaders of all the main Kurdish separatist movements met together in Paris for the first time in 60 years. Among issues of cultural identity and human rights, their main concern was that, the Iran–Iraq war having ended, the two protagonists might jointly turn their forces against the Kurds. This did not occur, but, particularly from 1990 onwards, there have been increasingly violent clashes between the Kurds and the Turkish army in southeast Anatolia. However, world opinion became once more focused upon the Kurdish issue after the defeat of Iraq during Operation Desert Storm. Counting on disarray in Baghdad, the Kurds of northeastern Iraq pressed for autonomy and promptly were attacked by the army. Many were killed and vast numbers fled into the mountains and the situation was only stabilized by United Nations' intervention.

The importance of Kurdistan is that it overlaps five states in a particularly sensitive part of the world. For almost the entire post-World War II period, there has been confrontation, if not outright conflict, between the Kurds and the three main host nations: Turkey, Iran and Iraq. Thus, in what is the world's major regional flashpoint, interstate tensions are heightened by the Kurdish issue.

Status

Kurdistan is a congenitally hostile regional environment. Iran fears Kurdish autonomy and attempts to play one faction against another. Tehran newspapers have gone as far as to insinuate that the West is interested in the creation of a 'new Israel'. Turkey continues to pursue Kurdish guerrillas while, at the same time, constructing in southeastern Anatolia, the Kurdish region, one of the world's major irrigation schemes. Control of the Euphrates by the Ataturk Dam to facilitate this development has already alienated both Syria and Iraq. However, Kurds in Iraq are dependent upon Turkey for their food and fuel lifeline and also for the vital air bases from which Western aircraft overfly the region and thereby constrain the Iraqi army. The ultimate fear must be that two or three of the main host nations join forces in an attempt to, once and for all, settle the Kurdish problem. The Kurds themselves are still faced with their age-old dilemma of whether to settle for autonomy or whether to push further for complete independence. Whichever they choose, Kurdistan will remain a key global flashpoint.

References

Anderson, E.W. and Rashidian, K. (1991), *Iraq and the Continuing Middle East Crisis*, Pinter, London.
Calvocoressi, P. (1991), *World Politics Since 1945*, Longman, London (Sixth Edition).
Eagleton, Jr, William (1963), *The Kurdish Republic of 1946*, Oxford University Press, Oxford.
Munroe, D. and Day, A.J. (1990), *A World Record of Major Conflict Areas*, Arnold, London.

38 The Kurile Islands

Description

The Kuriles form an island curtain, linking the most northerly of the main islands of Japan, Hokkaido, with the Kamchatka Peninsula (Russia). The islands divide the Sea of Okhotsk from the Pacific Ocean and have a population of between 20,000 and 25,000 civilians and perhaps as many as 10,000 Russian (formerly Soviet) troops.

The islands in dispute between Japan and Russia are located at the southern end of the chain and comprise three main islands, Etorofu, Kunashiri and Shikotan, together with the Habomai group. The areas of these are each respectively 3,139 km^2, 1,500 km^2, 255 km^2 and 102 km^2, giving a total of 4,966 km^2. Legally, all the Kurile islands are part of Russia, despite the fact that Suisho, one of the Habomais, is only 5 km from Hokkaido.

History and importance

The Treaty of Shimoda (or Treaty of Commerce, Navigation and Delimitation) in 1855, confirmed that the Kuriles south of Etorofu and including Etorofu itself, were Japanese, those to the north, Russian. This was further reinforced by the Treaty of St. Petersburg (1875) in which Japan also renounced claims to Sakhalin Island to the west. However, it is the view of Russia that, as a consequence of the Russo–Japanese war (1905) and the Portsmouth Treaty of the same year, because this was extracted under duress, the treaties of 1855 and 1875 have lost their validity.

In World War II, the Soviet Union and Japan signed a neutrality pact in April 1941. However this was before Germany's invasion of the Soviet Union or Japan's attack on the United States and on 8 August 1945 the Soviet Union declared war on Japan, which surrendered unconditionally on 14 August. Under the terms of the Yalta Agreement, made in February 1945 between the Soviet Union, the United States and Britain, the islands were given to the Soviets. However, Japan was not a party to the Agreement and, what is more, it was stated in both the Cairo Declaration (1943) and the Potsdam Declaration (1945) that the Allies were not fighting for territorial gain.

Thus, while the Russians consider the Yalta Agreement to represent the final answer to the case of the Kurile Islands, the Japanese consider the Agreement to be illegitimate and base their claims on the southern islands on the treaties of 1855 and 1875. On 8 September 1951, Japan signed the San Francisco Peace Treaty and, according to the Soviet Union, thereby gave up its claim to the disputed islands. In contrast, Japan insists that the Kurile Islands are not part of the northern territories which it gave up in 1951 and furthermore, the Soviet Union, having failed to sign the San Francisco Treaty, cannot pursue claims under it.

Finally, in 1956, under a Japan–Soviet Joint Declaration, the Soviets agreed to return Shikotan and the Habomais, but not until all foreign troops had been removed from Japan. The United States–Japan Mutual Cooperation Security Treaty, signed in January 1960, merely hardened the Soviet position.

The Kurile Islands, particularly those at the southern end of the chain, are of strategic and economic significance. Between them lie the choke points which control the movements of the Russian (Soviet) Pacific Fleet, based at Vladivostok and the ballistic-missile submarines which operate in the Sea of Okhotsk. They also provide access to one of the three most productive fishing grounds in the world. Indeed, on 10 December 1976, the Soviet Union declared a 200 nml fishing zone around the islands and on 26 January 1977 Japan responded in kind. By 27 May a fisheries agreement had been reached, but this was suspended in August 1978 when Japan signed a treaty of peace and friendship with China.

The intensity of feeling is illustrated by the fact that by 1977 more than 1,500 Japanese fishing vessels and almost 13,000 fishermen had been detained by

the Soviet Union. Between 1979 and May 1980, the Soviets increased their military build-up in the islands to some 13,000 troops but, since the collapse of the Soviet Union, the continuation of the military presence is uncertain.

Status

Much had been expected from the visit of President Gorbachev to Tokyo in April 1991, but, apart from reduction in troop numbers and aircraft and the admission that a dispute existed, little that was productive occurred. Mindful of the case of Alaska, Russia's President Boris Yeltsin has resisted the urge to make a deal. Despite the current state of the Russian economy, it would be politically difficult to make a straight land-for-cash swap, despite the desperate need for Japanese funds and technology which have been held up pending a resolution of the dispute favourable to Japan. The more likely process involves a gradual change of sovereignty, and since September 1991 relations have improved. On 14 October 1991 an agreement was reached to allow former Japanese and current Russian residents of the disputed Kuriles to have visa-free travel to Japan. This may presage future events.

References

Alexander, L.M. (1988), 'Choke Points of the World Ocean: A Geographic and Military Assessment' in E.M. Borgese, N. Ginsburg and J.R. Morgan (eds), *Ocean Yearbook* 7, University of Chicago Press, Chicago, pp. 340–55.

Boundary Bulletin, No. 1, (1991), International Boundaries Research Unit, Durham University.

Boundary Bulletin, No. 2, (1991), International Boundaries Research Unit, Durham University.

Boundary Bulletin, No. 3, (1992), International Boundaries Research Unit, Durham University, January.

Day, A.J. (ed.) (1984), *Border and Territorial Disputes*, Longman, London.

Downing, D. (1980), *An Atlas of Territorial and Border Disputes*, New English Library, London.

The Economist (1991) 20 April.

Falkenheim, P.L. (1987), 'Japan, the Soviet Union, and the Northern Territories: Prospects for Accommodation' in L.E. Grinter and Y.W. Kihl (eds), *East Asian Conflict Zones*, Macmillan, London, pp. 47–69.

Hara, K. (1991), 'Kuriles Quandary: The Soviet/Japanese Territorial Dispute' in *Boundary Bulletin*, No. 2, International Boundaries Research Unit, Durham University, pp. 14–16.

Olsen, E.A. (1987), 'Stability and Instability in the Sea of Japan' in L.E. Grinter and Y.W. Kihl (eds), *East Asian Conflict Zones*, Macmillan, London, pp. 70–96.

Prescott, J.R.V. (1985), *The Maritime Political Boundaries of the World*, Methuen, London.

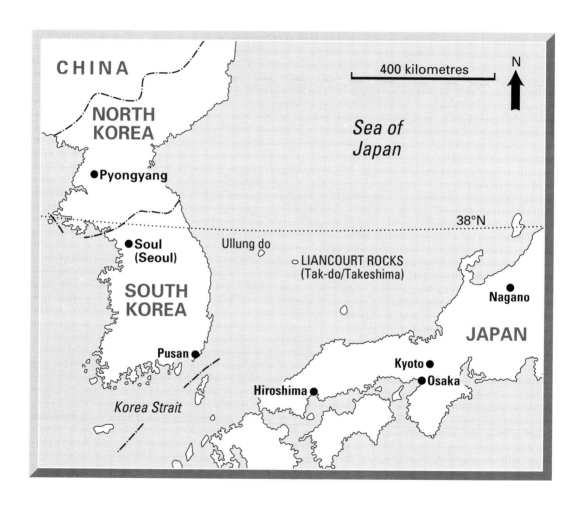

39 The Liancourt Rocks

Description

The Liancourt Rocks are a small island group known as Tak-do by South Korea and Takeshima by Japan. They lie in the southern part of the Sea of Japan some 210 km offshore from each disputant and therefore equidistant between them.

History and importance

The dispute over the Liancourt Rocks erupted between South Korea and Japan in 1952. Until that time, their potential importance had not been considered, but in 1954 South Korea occupied the rocks and subsequently refused international arbitration.

The rocks have strategic and economic importance and, as with so many similar disputes, there is an added symbolic significance. They are well located to provide surveillance over the entry to the Sea of Japan through the Korea Strait. Furthermore, since the Sea of Japan itself is one of the world's most strategic bodies of water, accessible to South Korea, North Korea, China and Japan, besides the two military superpowers, control over any centrally placed islands is important.

During the Cold War, the Sea was dominated by the Soviet Union, with its Pacific Fleet at Vladivostok, and the United States, with its bases in Japan. With the decline in East–West animosity, the situation is, if anything, more complex. If Japan develops its forces to become the pre-eminent power in the region, the Liancourt Rocks will be vital. If it fails to do so, other powers such as China or some other power may dominate, or there may be a vacuum. In any of these cases, the strategic location of the Liancourt Rocks is likely to be of great importance.

Economically, ownership of the Rocks would allow claims to some 16,600 nml^2 of sea and seabed. Not only are there important fisheries in the region, but the prospects for hydrocarbons look very promising.

Status

At present, the islands are held by South Korea and the issue is dormant. Future possibilities include joint development, similar to that at the southern end of the Korea Strait. Another option is for one side to withdraw its claim in return for a major discounting of the islands in the demarcation of maritime boundaries.

References

Downing, D. (1980), *An Atlas of Territorial and Border Disputes*, New English Library, London.

Olsen, E.A. (1987), 'Stability and Instability in the Sea of Japan' in L.E. Grinter and Y.W. Kihl (eds), *East Asian Conflict Zones*, Macmillan, London, pp. 70–96.

Prescott, J.R.V. (1985), *The Maritime Political Boundaries of the World*, Methuen, London.

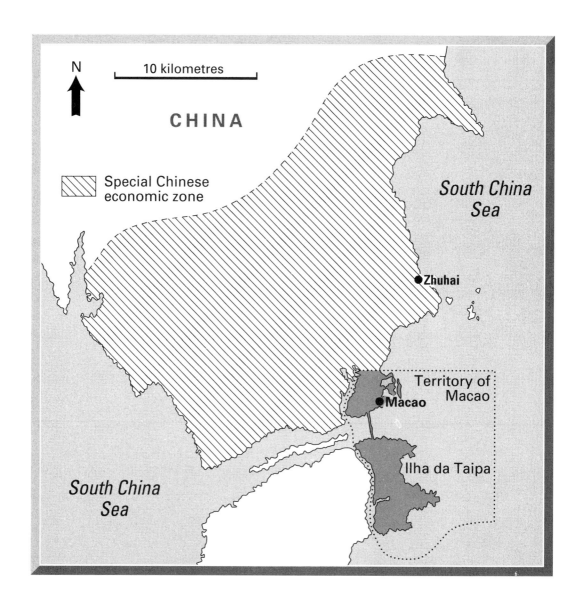

40 Macao

Description

The Portuguese overseas province of Macao is located on the South China Sea coastline of China. It is situated at the deltaic mouth of the Pearl River, some 105 km south of Canton and 65 km west of Hong Kong. The main city sits on a hilly peninsula, 5 km long, and is linked by a 200-m-wide isthmus to China. However, Macao also includes three small offshore islands: two Taipa Islands and Coloane Island, which are linked to the peninsula by a bridge and causeway. The total area is 15.5 km^2 and the population is estimated at approximately 380,000. Of these, only 4 per cent are Portuguese.

History and importance

Macao was founded and leased in 1557 when China agreed to a Portuguese settlement but not to sovereignty. It rapidly became the leading port for China's foreign trade, and several other colonial powers, particularly the Dutch, attempted to capture it. During the 17th century alone, the Dutch tried and failed to take Macao on four separate occasions. In 1586, Macao was declared a city and given the same status as the Portuguese city of Evora and in 1640, it was given the title of 'City of the Name of God, Macao, There is None More Loyal'.

In 1670, the Chinese confirmed Macao as a Portuguese possession. It continued to flourish until the 18th century when the rise of Hong Kong and the silting of the harbour led to the loss of its preeminent trading position. It then became more closely identified with smuggling and the establishment of gambling. In 1848–9, Portugal proclaimed Macao's complete separation from China, a position which was confirmed by the Manchu government in the Protocol of Lisbon (1887).

The importance of Macao to China is largely symbolic, for China would like to see the removal of the last vestiges of European colonialism. It remains an important outlet and large profits are made through its casinos, but its economy is far less significant than that of Hong Kong. The important factor for Portugal, which now exercises little influence in Macao, is to withdraw with dignity.

Status

Macao is a Special Territory of Portugal and the governor holds executive power for everything except foreign policy. There is a 17-member Legislative Assembly of whom 12 are elected, and the transfer of sovereignty to China has been agreed for 1999, with a 50-year transition period to follow. In 1990, a full Joint Working Committee was established, giving Beijing a voice in preparation for 1999. This paralleled the airport negotiations in Hong Kong.

However, on 28 September 1990, the governor resigned, following allegations of corruption and of mounting a smear campaign. He also made accusations concerning the conduct of representatives 'giving in' to Chinese pressure. His formula for defending Portuguese interests in Macao, which was similar to that being adopted in Hong Kong, included support for a massive development programme to secure Macao's economic autonomy and to involve private investors, thereby ensuring a perpetual Western influence. The major projects would be a second bridge to Taipa and an airport costing US$468 million (one of the three largest construction projects in Asia). While obviously China would benefit from such developments, the main concern of the potential superpower seems to be re-possession of the territory.

References

Boundary Bulletin, No. 1, (1991), International Boundaries Research Unit, Durham University.
Chambers World Gazetteer, (1988), Cambridge University Press, Cambridge.
Financial Times, (1992) 5 February.

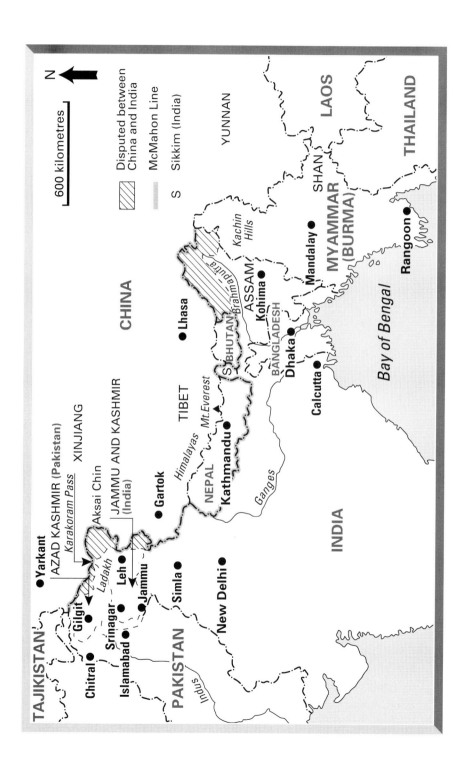

41 The McMahon Line

Description

At the Simla Conference (1913-4) on the status of Tibet, attended by Great Britain, China and Tibet, the McMahon Line was submitted by Sir Henry McMahon. It was proposed that the Line, which follows the watershed of the Himalayas, should be the boundary between Tibet and India. No agreement was reached and the entire length of the Sino–Indian frontier, some 14,000 km, is still in dispute. In fact, the proposal for the McMahon Line was initialled by both the Chinese and the Tibetan delegates but was repudiated by the Chinese government. A secret agreement was made with the Tibetan delegation to recognize the Line, but this was never ratified by the Tibetan government. Furthermore, the proposals failed to win the support of the Viceroy of India who stated that they should be regarded as 'personal and not at present carrying the endorsement of the Government of India'. Thus, the McMahon Line remains an idea but one which has resulted in conflict to the present day.

History and importance

In the eastern section, British control was extended to the McMahon Line in 1938, and in 1939 the Line appears as a frontier on official Survey of India maps. When Chinese troops occupied Tibet in 1950, Chinese maps showed a border extending to the Brahmaputra River in Assam. When in 1954 the Sino–Indian agreement on respect for territorial integrity and sovereignty, non-aggression and non-interference was signed, this was taken by India as an implicit guarantee for the existing frontier. Prime Minister Nehru said at that time of the McMahon Line: 'it is our frontier and we will not allow anyone to cross it.' In response, in 1956, China stated that although the McMahon Line was unfair, it would not be challenged in the interests of maintaining friendly relations. None the less, Chinese maps still showed the border far south of the line. Between 1956 and 1959 there followed a series of border incidents but no clashes between the armed forces of the two states.

However, in March 1959, with the uprising in Tibet, as a result of which some 12,000 Tibetans fled the country, relations deteriorated. By August, Chinese forces had occupied the Longju border post and the Indian garrison had been expelled. On 8 September, the Chinese Premier stated 'the Chinese government absolutely does not recognize the so-called McMahon Line'. Two days later, India replied that it 'stands firmly' on the McMahon Line. The area claimed by China south of the Line amounted to 83,000 km^2. Talks between India and China which followed failed, but in 1960-1 China dropped its claim to the Kachin hills area of northern Burma and accepted the Burmese part of the McMahon Line.

In late 1962, there were clashes near the tripoint of Bhutan, India and Tibet, and on 20 October the Chinese offensive at both ends of the McMahon Line began. Over the next month, troops advanced 160 km south of the Line at the western end and approximately 50 km at the eastern end. On 21 November, China announced a cease-fire and withdrawal to 20 km north of the Line where it was proposed that check-points should be established. In the event, the withdrawal took place but checkpoints were established at Dhola and Longju. Since then, there have been a number of clashes, but, despite a succession of moves to improve bilateral relations, no real progress has been made. Along the central section of the McMahon Line there have only been minor disagreements and the Nepal–China border was agreed in 1960-1.

In the western or Ladakh sector respect for current boundaries was included in Peace Treaties signed in 1684 and 1842. However, neither treaty defined where the boundary was and no survey was made until 1864. With independence in 1947, Jammu and Kashmir (including Ladakh) acceded to the Indian Union. Following the Chinese occupation of Tibet in 1950, the Ladakh frontier became part of the Sino-Indian border. Pre-independence British maps had consistently shown the boundary as following the crest of the Karakoram Range, but in July 1954 a new Indian government map showed the Kuenlun Crest as the frontier and the whole of Aksai Chin (36,000 km^2) as Indian. In fact, the area had been under Chinese control since 1950, the Aksai Chin being considered strategically important for China since it provided the link between Tibet and Xinjiang.

So isolated and rugged is this area that between 1956 and 1957 a high-altitude military road was built linking Tibet and Xinjiang, and India was unaware of its existence until alerted in a Chinese press report. During 1958 and 1959 there were various exchanges over the border but no concessions. On 27 November 1962, the Indian foreign ministry stated that Chinese troops had occupied some 15,500 km^2 of disputed area. This followed an offensive from 20 October when China had occupied all the Indian border posts east of the McMahon Line. Since then, despite occasional border incidents, the Aksai Chin has remained under Chinese control and the region has been generally peaceful.

On 3 May 1962, China and Pakistan announced an agreement to demarcate the common border in Kashmir, but India strongly protested stating that Kashmir was 'an integral part of the Indian Union'. However, the delimitation agreement was signed on 2 March 1962. Of the 8,800 km^2 in dispute, Pakistan obtained approximately 3,300 km^2 and China 5,500 km^2. Pakistan also abandoned a claim to some 33,000 km^2 of Xinjiang territory.

Thus the China–Pakistan border has been agreed but that between China and India, involving an area of 119,000 km^2, is still in dispute. Negotiations in 1981 and 1982 between China and India made no progress but in the latter year, a broadcast on Beijing radio stressed peaceful coexistence and the maintenance of the status quo: 'the Chinese government, although not recognizing the McMahon Line, does not cross this Line for the sake of striving for a negotiated solution'.

Status

Although never accepted in its entirety, the McMahon Line is at present relatively stable and peaceful. The sections between Burma and China, India and Bhutan and India and Nepal are not in dispute. That between China and Pakistan has been the subject of peaceful negotiations. It is the Sino-Indian section of the McMahon Line which is subject to periodic pressure from each side. It constitutes a constant potential irritant and occasionally becomes a live issue. It is unlikely ever to become a major conflict zone, but it remains an underlying factor in bilateral relations.

References

Boyd, A. (1991), *An Atlas of World Affairs*, Routledge, London.

Bradnock, R.W. (1990), 'South Asia: The Frontiers of Uncertainty' in N. Beschorner, St.-J.B. Gould and K. McLachlan (eds), *Sovereignty, Territoriality and International Boundaries in South Asia, South West Asia and the Mediterranean Basin*, Proceedings of a seminar held at the School of Oriental and African Studies, University of London, pp. 1–8.

Day, A.J. (ed.) (1984), *Border and Territorial Disputes*, Longman, London.

42 The Magellan Strait

Description

The Magellan Strait is the longer but wider and less tortuous inter-oceanic channel linking the Atlantic and Pacific Oceans which provides an alternative to the Beagle Channel in the far south of South America. It lies on the northern side of Tierra del Fuego and, apart from its entrance, is wholly within the territory of Chile. It is navigable by comparatively large ships and is sheltered so that it provides a more rapid and safer route than that round Cape Horn. Therefore, it is of significant strategic importance.

History and importance

From 1847, when it protested over the foundation by Chile of Fuerte Bulnes (1843) on the shores of the Magellan Strait, Argentina has claimed historical and judicial rights to the Magellan inter-oceanic connection between the Atlantic and the Pacific. The dispute was ostensibly settled by the Boundary Treaty of 23 July 1881, by which Chile maintained control over the Magellan Strait, but the Strait itself remained neutral for shipping.

Since the territorial sea, fronting the eastern entrance to the Strait was limited to 3 nml, the Strait posed no particular problems until relatively recently. However, with the expansion of state jurisdiction, the stakes rose considerably with both Argentina and Chile claiming 12 nml territorial seas and 200 nml EEZ, fronting the eastern entrance to the Strait. It should be noted that a Chilean EEZ of 200 nml would almost reach the EEZ of the Falkland Islands. As indicated in the discussion on the Beagle Channel (Map 12), during 1978 it proved impossible to reach agreement on a number of issues, including the Magellan Strait, and conflict between Chile and Argentina seemed likely by the end of that year. However, intervention by the Papacy defused the situation.

Apart from the obvious aspect of offshore boundaries and thereby control over resources, the Magellan Strait must also be seen in its geopolitical context. There is a strong Argentinian perception that Patagonia (the southern part of mainland Argentina), the southern islands of South America and of Antarctica are threatened by Chilean ownership of territory on the Atlantic seaboard. Chilean policy, on the other hand, is governed by an absolute desire to keep Argentina out of the Pacific. The Magellan Strait has been useful to Chile in keeping the dividing line between the two countries as far east as possible. Thus, the Strait dispute is an aspect of the Chilean and Argentinian bi-oceanic principle: Chile in the Pacific, Argentina in the Atlantic.

Status

The issue of the Magellan Strait has been complicated by the 'Beagle Channel dispute', since any resolution of the latter, alone, would have led to the re-opening of the former. A comprehensive, simultaneous approach to both disputes was therefore necessary and this contributed to the complex and lengthy negotiations for the Peace and Friendship Treaty 1985. While it is impossible to state that this potential flashpoint will remain acquiescent, at the moment both states are more preoccupied with internal developments than with threatening each other.

References

Child, J. (1985), *Geopolitics and Conflict in South America*, Praeger/Hoover Institution Press, Stanford.

Morris, M.A. (1986), 'E.E.Z. Policy in South America's Southern Cone' in E.M. Borgese and N. Ginsburg (eds), *Ocean Yearbook* 6, University of Chicago Press, Chicago, pp. 417–37.

Morris, M.A. (1988), 'South American Antarctic Policies' in E.M. Borgese and N. Ginsburg (eds), *Ocean Yearbook* 7, University of Chicago Press, Chicago, pp. 356–71.

Santis-Arenas, H. (1989), 'The Nature of Maritime

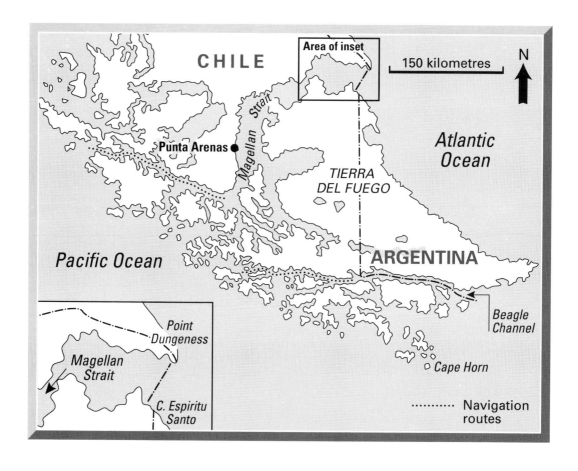

Boundary Conflict Resolution between Argentina and Chile, 1984' in *International Boundaries and Boundary Conflict Resolution*, 1989 Conference Proceedings, International Boundaries Research Unit, University of Durham, pp. 301–22.

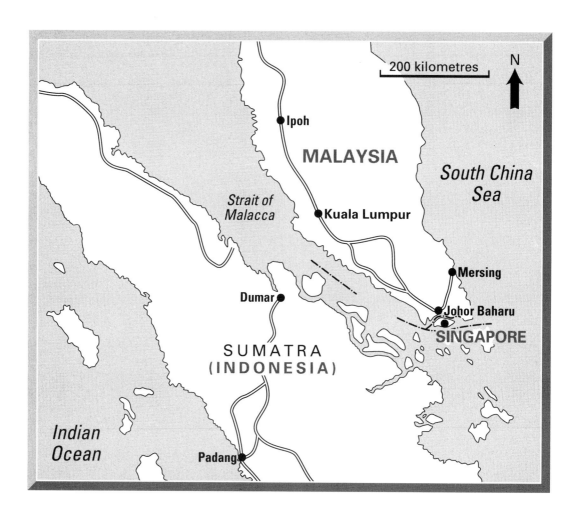

43 The Strait of Malacca

Description

Situated between the Malay Peninsula and Sumatra, the Malacca Strait, one of the most important choke points in the world, links the South China Sea and Pacific with the Indian Ocean. The Strait is 805 km long and varies in width from 320 to 50 km. However, towards its southern end, the channel is only some 12 km wide. Not only is navigation hindered by islands, but there are numerous fishing boats, ferries and wrecks. According to different authorities, the number of daily ship counts averages anything from 230 to 600. Using those figures, the only choke point to rival the Strait of Malacca is Dover Strait.

History and importance

Portugal captured control of the Strait in 1513, but there were constant attacks from local powers, particularly the Malay Sultanates. Eventually, in 1641, the Portuguese were evicted by the Dutch. In 1824, British merchants founded Singapore at the eastern entrance to the Strait and the whole Malay Peninsula came under British protection by the end of the 19th century.

During World War II, the Japanese captured the Strait, but since that time control has been left to the three riparian states: Malaysia, Singapore and Indonesia. Concern over superpower rivalry, safety of navigation and possible pollution has led to trilateral cooperation.

In August 1957, Indonesia declared a 12 nml territorial sea limit and in August 1969 Malaysia followed suit. Thus, the southern end of the Strait was completely territorial sea and, as a result, there was international concern over the freedom of navigation and the access of foreign navies.

On 16 November 1971, at a meeting in Kuala Lumpur, the three states confirmed that they were solely responsible for the safety of navigation and furthermore that such safety and the internationalization of the Strait should be considered as separate issues. Indeed, Indonesia and Malaysia agreed that the Strait was not international, although they recognized 'innocent passage' rights. Singapore expressed reservations about this decision. The resulting Kuala Lumpur Declaration provoked strong international protest and later Indonesia and Malaysia declared that tankers of 200,000 tonnes and warships were prohibited from transiting the Strait. Such ships were therefore required to take the more circuitous route through the Lombok Strait.

The issue was defused in 1974 by the start of the United Nations Conference on the Law of the Sea (UNCLOS) III negotiations which, by 1982, had reached broad agreement on transit passage through international straits. However, many problems of definition remained, for instance, concerning 'continuous and expeditious transit' and 'straits used for international navigation'. Moreover, transit passage was not deemed to allow 'research or survey activities', although these are carried out by a wide range of ships, including warships.

No mention was made of 'submerged passage'. Given the intense rivalry in the region between the Soviet Union and the United States, this last point was of particular significance, although its importance may now have declined.

Therefore, the Strait of Malacca is, commercially, one of the world's major sea-lanes and, strategically, a key choke point. It is, in almost all cases, the most cost-effective route for transit between the Pacific and Indian Oceans, the alternative straits of Sunda, Lombok and Ombia-Wetar all adding significantly to distance. For the most important route, that between Yokohama and the Persian/Arab Gulf, Sunda adds 9 per cent. Lombok 18 per cent and Ombia-Wetar 23 per cent. However, Very Large Cargo Carriers (VLCCs) of 200,000 dwt or over usually transit the deeper Lombok Strait, returning through Malacca.

The most vital transit trade through the region

is oil from the Gulf, destined for Japan. In 1990, two-thirds of Japan's oil imports originated in the Persian/Arabian Gulf region and was shipped via Indonesian waters. Strategically, the Strait has played an integral role in superpower strategy, but it is also vital in the protection of the interests of the riparian states and particularly the port economy of Singapore.

Status

With the emerging economics of Association of South East Asian Nations (ASEAN) and the increasing geopolitical status of Japan, the Strait is likely to increase in both economic and strategic importance. There are, however, a number of major potential problems. With the proliferation of peace zones and particularly the Zone of Peace, Freedom and Neutrality (ZOPFAN), which includes Malacca, the concept of 'innocent passage' is under pressure. Nuclear-powered ships and those carrying nuclear weapons could have particular problems. There are also concerns over the extension of the military roles of both China and Japan.

Over the past decade, the Chinese South Sea Fleet has increased from 20 to 70 surface vessels, and China also possesses the third-largest submarine fleet in the world. There could also be potential problems over the rise of an Indian naval presence, possibly in relation to the construction of the 'export processing zone' on the Nicobar Islands, 800 km from India, but only 80 km from Sumatra. Indeed, the Nicobar Islands are well located to control the western approaches to the Malacca Strait and the initiative is thought by many to be a cover for the construction of military airports and naval facilities. At present, the Indian Navy is the largest in the Indian Ocean and the seventh largest in the world.

In contrast, the Russian Pacific Fleet has run down its presence and pulled out of the Indian Ocean and the Mediterranean for the first time in 30 years, and the United States, through the projected loss of its base in the Philippines, will have a reduced presence. Thus, given the huge military and strategic changes in the Pacific and Indian Ocean theatres, it is no exaggeration to suggest that it will remain a key potential flashpoint.

References

Alexander, L.M. (1988), 'Choke Points of the World Ocean: A Geographic and Military Assessment' in E.M. Borgese, N. Ginsburg and J.R. Morgan (eds), *Ocean Yearbook* 7, University of Chicago Press, Chicago, pp. 340–55.

Chambers World Gazeteer (1988), Cambridge University Press, Cambridge.

Cottrell, A.J. and Hahn, W.F., *Naval Race or Arms Control in the Indian Ocean*, Agenda Paper No. 8, National Strategy Information Center, Inc., New York.

Cunha, D.D. (1991), 'Major Asian Powers and the Development of the Singaporean and Malaysian Armed Forces', *Contemporary S.E. Asia*, 13, No. 1, June, pp. 55–71.

Leng, L.Y. (1982), *Southeast Asia: Essays in Political Geography*, Singapore University Press, Singapore.

Leng, L.Y. (1987), 'Access to S.E. Asian Waters by Naval Powers', *Contemporary S.E. Asia*, 9, No. 3, December, pp. 208–20.

Naidu, G.V.C. (1991), 'The Indian Navy and S.E. Asia', *Contemporary S.E. Asia*, 13, No. 1, June, pp. 72–85.

Rais, R.B. (1986), *The Indian Ocean and the Superpowers*, Croom Helm, London.

Weatherbee, D.E. (1987), 'The South China Sea: From Zone of Conflict to Zone of Peace?' in L.E. Grinter and Y.W. Kihl (eds), *East Asian Conflict Zones*, Macmillan, London.

Stubbs, R. (1991), 'Malaysian Defence Policy', *Contemporary S.E. Asia*, 13, No. 1, June, pp. 44–56.

44 Mayotte Island

Description

Mayotte Island (12°50S, 45°10E) lies at the northern end of the Mozambique Channel, which runs for some 900 nml between the island nation of Mozambique and East Africa; at its narrowest, it is 230 nml wide. Throughout the channel there are island groups: the Comoros, the Aldabra Group within the Seychelles Islands and the French territories of Mayotte, Îles Glorieuses, Juan de Nova, Bassas da India and Île Europa.

The Comoros, comprising four main islands: Njazidja, Nzwani, Mwali (Grande Comore, Anjouan, Moheli) and Mayotte (Mahore) and many small islands lie between 300 and 500 km northwest of Madagascar. The total land area is 2,236 km^2 and the population, which is 99.7 per cent Muslim, totals 370,000 (1989). Mayotte itself has an area of 375 km^2 and a population of 67,000 (1989).

History and importance

France, which had been the colonial power since the late-19th century, granted independence to Madagascar in 1960 and the Comoros in 1975, but the people of Mayotte insisted on retaining French protection. In 1958 the Mouvement Populaire Mahorais (MPM) was founded to safeguard the interests of Mayotte. On 10 September 1972, in the Comoros Chamber of Deputies, the vote was taken to promote independence, in friendship and cooperation with France. This was opposed by the five MPM members.

On 15 June 1973, a Joint French–Comoro Declaration stated that independence would be attained within five years. On 22 December in the following year, a referendum resulted in a 96 per cent vote in favour of independence, but in Mayotte, there was a 63 per cent vote against. On 26 June 1975, the French parliament passed the Comoro Islands Independence Bill, which required the establishment of a constitutional committee and island-by-island referendum. This was deemed unacceptable by the Comoro government and the Chamber of Deputies voted in favour of immediate independence on 6 July. The MPM delegates, who abstained from voting, telegraphed the French president, placing Mayotte 'under the Protection of the French Republic'. As a result, on 10 July, France stated that it would grant independence but not necessarily as one unitary state to the Comoros and later that month French troops left Grande Comore, but 200 foreign Legionnaires remained on Mayotte.

On 10 December, the French National Assembly passed a bill, recognizing the independence of the three Comoro Islands and providing for a referendum on Mayotte to determine the island's future. A further provision was that if the decision were to be in favour of remaining part of the French Republic, there should be a second referendum two months later to decide the status of Mayotte as either an overseas department or a territory.

On 8 February 1976, the first referendum produced a 99.4 per cent majority in favour of remaining with France, and France fully supported this stand for self-determination. Immediately prior to the referendum, the United Nations Security Council met and issued a draft resolution declaring that the referendum was an interference in the internal affairs of the Comoros and that talks should be held on a hand-over. This was vetoed by France.

Later that year, on 21 October, the United Nations General Assembly passed a motion by 102 votes to 1, calling on France to withdraw from Mayotte. On 6 December, a further resolution reaffirming the Comoro sovereignty over Mayotte and calling for early negotiations, was passed by 112 votes to 1. This was rejected by France as 'an impermissible interference in the internal affairs of France'. On 22 December 1979, the French Senate and the National Assembly promulgated a law, stating that Mayotte was 'part of the French Republic and cannot cease to belong to it without the consent of its population'.

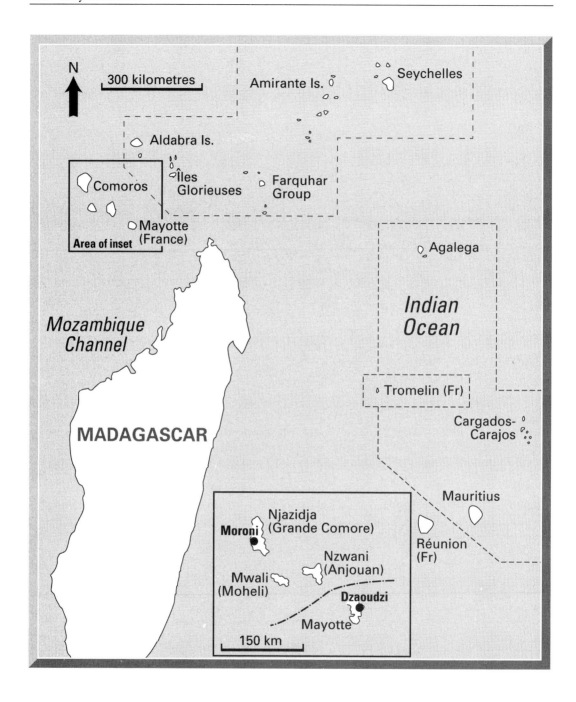

Mayotte occupies a strategic position in the Mozambique Channel, which is itself a choke point on the Cape Route from the Persian/Arabian Gulf to Europe. This is the main route used by super-tankers, which are too large to pass through the Suez Canal. Furthermore, some of the best fishing grounds in the Indian Ocean lie between Madagascar and Mayotte.

Politically, Mayotte presents a classic case, like the Falklands and Gibraltar, of the last vestiges of empire. To the Third World, such persisting territories represent a failure of the decolonization process and even evidence of neo-imperialism. Another view is that they are simply examples of the right to self-determination. In all three cases, almost the total population is in favour of retaining the current status.

Status

Mayotte continues under its chosen system of French occupation. However, problems of the delimitation of the EEZ and of fishing and maritime boundaries loom. The Comoros and Madagascar are both strongly opposed to Mayotte's current status and discussion could quickly dissolve into conflict. However, Mayotte sends one deputy to the French National Assembly and is an official department of France, or more correctly a 'territorial collectivity' (collectivité territoriale). Therefore, like the Falkland Islands with the British, it would expect the full support of the French military in the event of problems.

References

Blake, G.H. and Anderson, E.W. (1982), *International Maritime Boundary Delimitation and Islands: Examples Analogous to Libya–Malta*, University of Durham (unpublished), September.

Day, A.J. (ed.) (1984), *Border and Territorial Disputes*, Longman, London.

Downing, D. (1980), *An Atlas of Territorial and Border Disputes*, New English Library, London.

The Economist Atlas (1989), Economist Books, Hutchinson, London.

Prescott, J.R.V. (1985), *The Maritime Political Boundaries of the World*, Methuen, London.

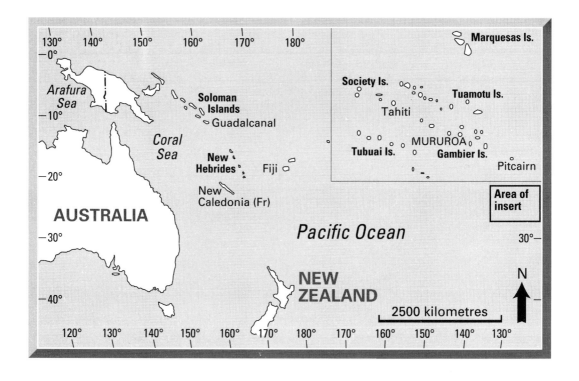

45 Mururoa Atoll

Description

Mururoa Atoll is one of the Tuamotu Islands, one of five island groups that comprise French Polynesia about midway between South America and Australia. French Polynesia comprises some 120 defined islands, a land area of some 4,000 km^2 and a total population of 160,000.

History and importance

One of the key political issues for the South Pacific states is that their region is the only one in the world where nuclear tests are carried out by an extra-regional power. The post-war period has been dominated by this factor and by moves towards cooperation and independence. In 1947, the South Pacific Commission (SPC) was founded by the relevant metropolitan countries: Australia, Britain, France, the Netherlands, New Zealand and the United States. Until 1974, it enjoyed only an advisory role, but from then on it became the governing body. In 1971, the South Pacific Forum was founded as a symbol and instrument of regional decision-making.

In 1966, the first atmospheric tests took place on Mururoa, to be followed by the first underground nuclear test in 1975. In 1985, the Treaty of Rarotonga established the South Pacific nuclear-free zone, recognized and respected by the Soviet Union and China, but not by France, Britain and the United States. In the same year, the *Rainbow Warrior*, a ship which the environmental group, Greenpeace, had planned to take to impede explosions at Mururoa, was sunk in Auckland Harbour by French secret agents, and world attention was again focused on French nuclear testing.

In September 1985, French President Mitterrand visited Mururoa to explain the French point of view and in the same month, Australian Prime Minister Robert Hawke approached the United States to put pressure on France over Mururoa. However, the United States re-confirmed its long-held view of the strategic primacy of the North Atlantic over the South Pacific and stated that the French decision was vital to the modernization of its nuclear deterrent.

However, Australia continued to protest, particularly after the French agents convicted of the *Rainbow Warrior* sinking were returned to France, contrary to the United Nations secretary-general's arbitration, which required that they be incarcerated locally. Australia stated that 'if France insisted on conducting these tests, it should do so on its home territory, especially if the tests were as harmless as France claimed'. France quoted a report produced by scientists from Australia, New Zealand and Papua New Guinea (1938), which asserted that the nuclear tests were completely harmless. However, as Australia had pointed out, the report had not cleared the tests of long-term environmental consequences.

Mururoa was originally selected for tests because of its isolation, which allegedly made it a particularly suitable site for atmospheric testing. Furthermore, the structure of the island made it a convenient but not essential site for underground tests.

Status

French Polynesia is classified as a French overseas territory and it sends two deputies to the French National Assembly. Within the islands there are wide-ranging social and economic problems, allied to a fear of the greenhouse effect and rising sea levels. A majority of the population favours more autonomy, but a growing minority supports full independence. However, France is unlikely to give up the islands, which are essential to support its global reach and to sustain its perception as a world power. There is also the huge potential of the EEZ delimitation.

With regard to the nuclear testing, the French attitude is that the weapons and the territory, and therefore the decision, are all French. In particular, France objects to the protests of Australia and New Zealand. Mururoa may be 7,100 km from Canberra, but Paris is only 5,200 km from the Russian test site at Semipalatinsk, and Alice Springs is nearer to the Chinese test site at Lob Nor than to Mururoa.

The regional position remains strongly opposed to testing. If the tests are harmful, they should be terminated; and if they cause no harm, they should be conducted in metropolitan France. The deadlock over this issue is likely to continue, particularly in the light of French aspirations to lead a European defence organization.

References

Downing, D. (1980), *An Atlas of Territorial and Border Disputes*, New English Library, London.

Dorrance, J.C. (1990), 'U.S. Strategic and Security Interests and Objectives in Australia, New Zealand, and the Pacific Islands' in Dorrance *et al.*, *The South Pacific: Emerging Security Issues and U.S. Policy*, Institute for Foreign Policy Analysis Inc., Cambridge, Mass. and Washington DC, pp. 1–26.

Thakur, R. (1990), 'Nuclear Issues in the South Pacific' in Dorrance *et al*, *The South Pacific: Emerging Security Issues and U.S. Policy*, Institute for Foreign Policy Analysis Inc., Cambridge, Mass. and Washington DC, pp. 27–51.

46 Nagorno-Karabakh

Description

Nagorno-Karabakh is an autonomous enclave, populated by some 180,000 Christian Armenians, situated in Muslim Azerbaijan. Ancestral hatred between the Armenians and Azeris has existed for over one thousand years and, while it reappeared at various times over the centuries, it has been able to reassert itself most strongly following President Gorbachev's *glasnost* policy.

Armenia had been the first state formally to adopt Christianity as its official religion (301), but in 387, it was partitioned between the empires of Persia (Zoroastrian) and Byzantium (Christian). There followed a succession of successes and setbacks as the Persians, the Turks and, later, the Russians struggled for control of the Transcaucasus. In 1905, during the first Russian revolution, there were clashes between Armenians and Azuris and, in 1915, some 600,000 Armenians were massacred by the Turks.

In 1918, the Armenians joined with the Russians to suppress a Muslim revolt, but later in the same year, the Azuris, under Turkish protection, exacted revenge. In March 1918, under the Treaty of Brest–Litovsk, Russian Armenia became an independent republic under German control, and one month later Azerbaijan declared its independence.

In 1919, following conflict over the enclave, the British military administration awarded Nagorno-Karabakh to Azerbaijan. One year later, the Treaty of Sèvres established Greater Armenia as an independent state, but in the same year, independence ended when Soviets and Turks, together, recaptured the Caucasus. Both Azerbaijan and Armenia were then proclaimed Soviet Socialist Republics and the Communist leadership of Azerbaijan passed Nagorno-Karabakh to Armenia. However, this was reversed in 1923 when Stalin designated the enclave an Autonomous Region and awarded it back to Azerbaijan.

In 1924, Nakhichevan, populated by Azuris, but separated from Azerbaijan by an area of Armenia, was established as an Autonomous Republic and attached to Azerbaijan. This further illustrated the problem of attempting to draw boundaries between the Armenian and Azerbaijan Soviet Socialist Republics. Not only were the two enclaves established, but many Armenians continued to live in Azerbaijan and Azuris in Armenia.

With *glasnost*, in 1985, Armenians demanded the incorporation of Nagorno-Karabakh into their republic and by 1988, there had been serious clashes between Armenians and Azeris over the issue. It is estimated that by the end of 1988, some 160,000 Armenians had left Azerbaijan and only a slightly lower number of Azeris had left Armenia.

As a result of the problems, in January 1989, Nagorno-Karabakh was placed under direct rule from Moscow and in October of the same year, the Azerbaijani Popular Front and the Armenia Pan-national Movement were both legalized. On 28 November 1989, Moscow ended the direct rule of Nagorno-Karabakh and on 1 December, Armenia declared the enclave part of the Armenian Republic. However, before the end of the year, Azerbaijan had virtually cut off communication links to Nagorno-Karabakh and there was a report that the water supply for Stepanakert had been deliberately polluted. In January 1990, Nakhichevan declared independence and the Baltic Council offered to mediate between representatives of the warring factions.

Nagorno-Karabakh has achieved a high public profile as an example of a severely ostracized enclave. Geopolitically, it represents an attempt to rationalize by political boundaries a complex ethnic mosaic. Furthermore, the Azuris are divided by the Russian–Iranian border and Azerbaijan is crucial for its oilfields, centred on Baku.

Status

Despite the signing of the Zheleznovodsk Com-

muniqué on 23 September 1991, attacks on Nagorno-Karabakh by the armed forces of Azerbaijan have increased. It had been intended that the peace would be maintained by the Soviet military, but in fact the forces withdrew and only some 1,200 soldiers of the Soviet Fourth Army remained in Stepanakert.

In December, the Armenian authorities of the enclave declared independence and elections were held. However, the Azuris within the enclave refused to take part and clashes broke out again, after the president of Azerbaijan had, on 2 January 1992, signed a decree, introducing direct presidential rule over Nagorno-Karabakh and the Azuri–Armenian borderlands. Throughout January 1992, clashes continued and the conflict has so far claimed some 1,000 lives.

There have been international calls for a United Nations Peace-keeping Force to be introduced, but all diplomatic pressure and humanitarian aid is conditional upon the fulfilment of human-rights criteria, itself dependent upon the implementation of the Zheleznovodsk Communiqué. As the states of the former Soviet Union continue to organize themselves and their relationships with each other and, particularly with Russia, Nagorno-Karabakh seems likely to remain a key geopolitical issue.

References

Boundary Bulletin, No. 1, (1991), International Boundaries Research Unit, Durham University.

Boundary Bulletin, No. 3, (1992), International Boundaries Research Unit, Durham University, January.

Glezer, O., Kolosov, V., Petrov, N., Streletsky, V., and Treyvish A. (1991), 'A Map of Unrest in the Soviet Union' in *Boundary Bulletin*, No. 2, International Boundaries Research Unit, Durham University, pp. 16–20.

Munro, D. and Day, A.J. (1990), *A World Record of Major Conflict Areas*, Arnold, London.

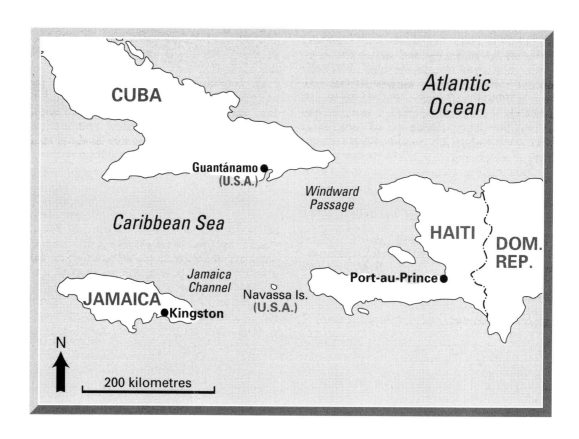

47 Navassa Island

Description

Lying between Jamaica and Haiti to the south of the Windward Passage, Navassa Island is centrally situated in the Jamaica Channel. It lies 50 km west of Haiti and 150 km east of Jamaica and has an area of 5 km^2. The population is transitory, comprising only tourists during the holiday season, but there are regular 'flag-waving' visits by the United States Coast Guard.

History and importance

Under an Act of Congress of 1860, the United States claims Navassa as a 'guano island' (that is, a source of nitrate fertilizer from natural deposits of bird droppings) and during World War II, a lighthouse was constructed there. During the 1950s, Haiti, which also claims the island, built a church for passing fishermen. However, the issue rose to prominence on 18 July 1981 when a group of six Haitians, having refused to seek landing permission from the United States, landed by government helicopter and established symbolic occupation of the island. A crew from the Haitian national television was on hand and the Haitian Communications Authority allocated the island a Haitian radio-call prefix (as opposed to an American prefix). However, the occupiers were arrested by six American marines and flown to a United States naval vessel.

Apart from its strategic location, Navassa is important to both claimants as it would allow a territorial claim of some 4,100 nml^2 of the surrounding sea.

Status

If sea-bed resources are discovered, Navassa could become a significant Caribbean flashpoint. Indeed, since Cuba and Haiti have already agreed to a maritime boundary which cuts into the Navassa equidistance area, there are already signs of potential conflict between the nations involved.

References

Day, A.J. (ed.) (1984), *Border and Territorial Disputes*, Longman, London.
Prescott, J.R.V. (1985), *The Maritime Political Boundaries of the World*, Methuen, London.

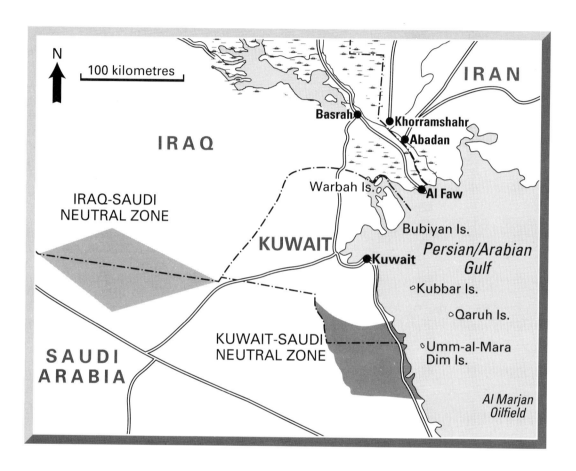

48 The Neutral Zones

Description

In the northern part of the Arabian Peninsula, two neutral zones are recognized, one to the west and the other to the south of Kuwait. The former, which occupies a diamond-shaped area on the Iraq–Saudi Arabia border, has an area of 6,500 km^2. Its eastern side begins at the extremity of the western frontier of Kuwait at the junction of wadis al-Audja and al-Batin. The northern border with Iraq extends for 190 km and the southern, with Saudi Arabia, for 200 km. The north–south width of the zone is 65 km.

The southern Neutral Zone consists of 64 km of coastline and its inland extension of some 75 km, abutting directly on to the southern boundary of Kuwait. The approximate area is the same as that of the western Neutral Zone.

History and importance

The Iraq–Saudi Arabia Neutral Zone results from bitter rivalry between the two countries. On 5 May 1922, under British aegis, the Treaty of Mohammerah (Khorramshahr) was signed by Ibn Saud and the British High Commissioner for Iraq. However, the boundary was not defined until 2 December, when, under the Uqair Convention, the Neutral Zone was also established. The agreement on the administration of the Neutral Zone was not signed until May 1938. In July 1975, it was reported in Saudi Arabia that an agreement on an equal division of the Neutral Zone had been achieved, but this remained unratified. On 26 December 1981, an agreement was reached which defined and 'fixed' the border and provided for the 'division of the Neutral Zone'. As a result, the frontier was 'stabilized'.

The Kuwait–Saudi Arabian Neutral Zone was also established under the Uqair Convention of 2 December 1922. In 1960, an equal division was agreed in principle and on 7 July 1965, the Agreement was signed at al-Hadda. On 18 December 1969, the instruments of ratification were exchanged. The one issue remaining concerns the offshore islands of Kubbar, Qaruh and Umm al-Maradim, located some 20 nml offshore. In 1961, Kuwait offered to share oil proceeds from the Neutral Zone in exchange for Saudi Arabian acknowledgement of Kuwait's sovereignty over the islands. Saudi Arabia refused. Equidistance lines would give the three islands approximately 2,000 nml^2 of sea space.

While the significance of the Iraq–Saudi Arabian Neutral Zone concerns border integrity and tribal movements, that of the Kuwait–Saudi Arabia Neutral Zone involves oil production. This Zone contains 0.5 per cent of the world's total oil reserves and has a resources-to-production ratio of 50. In 1990, production was 14,300 million tonnes or 275,000 bpd. This constituted 0.5 per cent of the world production total, a figure approximately one-quarter less than that produced in 1989.

Status

Officially, both boundary issues have been resolved and the outstanding problem concerns only the offshore islands of the Kuwait–Saudi Arabia Neutral Zone. However, following the official border settlement of 1966, Kuwait and Saudi Arabia continue to dispute the oil revenues of their Neutral Zone. During the Iran–Iraq War, they subsidized Iraq with profits from the sale of the Zone's oil. It is now assumed that there is a peaceful sharing of revenues between the two states, but offshore discoveries could yet lead to further disharmony.

With the ending of the Gulf Conflict (1991), the boundary between Kuwait and Iraq is being redefined, but there is no evidence that the status of the Iraq–Saudi Arabia Neutral Zone is affected. Thus, it is more the Kuwait–Iraq boundary than either Neutral Zone which is likely to remain a flashpoint.

References

Day, A.J. (ed.) (1984), *Border and Territorial Disputes*, Longman, London.
The Economist (1989) 21 January.
Prescott, J.R.V. (1985), *The Maritime Political Boundaries of the World*, Methuen, London.

49 Northern Ireland

Description

The British Isles comprise two main islands, Britain and Ireland. Britain is composed of England, Scotland and Wales and Ireland of the Republic of Ireland and Northern Ireland. England, Scotland, Wales and Northern Ireland together are one state, the United Kingdom (UK). Thus, Ireland is divided between two independent states and the boundary between them is an international border. However, the border does not coincide with the original fourfold division of Ireland into provinces. The northern province, Ulster, was divided into nine counties, but only six of these (Derry, Antrim, Down, Armagh, Fermanagh and Tyrone) are included in Northern Ireland. The area of these six is 14,120 km^2 and the population approximately 1.6 million. Although it comprises only some 6 per cent of the territory and under 3 per cent of the population of the United Kingdom, Northern Ireland has provided the focus of internal geopolitical attention for the past 30 years.

There are three fundamental cleavages in the society of Northern Ireland. The most important is that between Roman Catholics and Protestants, the breakdown according to religion being: Roman Catholic 28 per cent, Presbyterian 23 per cent, Church of Ireland 19 per cent and Methodists 4 per cent. A second division is between those who support the unification of Ireland and those who wish to remain within the United Kingdom. The former include the various shades of Republican and the latter the Unionists and the Loyalists. The other cleavage is the normal one, based on wealth. Although these three divisions by no means coincide, natural inclination and particularly the pressure of events has produced two opposing camps, each labelled according to religion. Broadly speaking, the Roman Catholics comprise the poorer elements of society and those who support unification, whereas the Protestants include the wealthier people and as a group wish to retain strong UK ties. The least discriminating of these is according to wealth since a significant proportion of the Protestant population is poor. Since the Protestants are in the majority and the political parties, with one exception, reflect the deep division of society, power sharing has not proved viable.

Apart from the structure of society, the other major components of the problem are the history of Anglo–Irish relations and the status of the issue. For at least 800 years Ireland has been dominated by its far more powerful neighbour and, even since it became a fully independent republic in 1937, the country has lived in the shadow of the United Kingdom. This unequal relationship has, until the emergence of the European Community, influenced dealings between the two countries. It has also affected the status of the Northern Ireland question. The United Kingdom has consistently viewed the issue as an internal matter, whereas, for the Republic of Ireland it is an international problem. At the least, given the operations of the Irish Republican Army (IRA), it must be conceded that there are international dimensions.

History and importance

The intractable nature of the Northern Ireland problem can only be appreciated in the context of the complex historical relationships between Britain and Ireland. By 1172, Henry II of England had conquered Ireland and had instigated the movement of settlers, at this time Anglo-Normans, into the country. The differences between the new settlers and the original population were reinforced in 1366 when the statutes of Kilkenny forbade intermarriage between English and Irish. Later, in 1541, Henry VIII adopted the title of King of Ireland and, during the Reformation which followed, Roman Catholics were persecuted. In the period of the counter-Reformation, two major insurrections were crushed by the English and in 1607, James I initiated the plantation of

Ulster with English and Scottish Protestants. Thus, the foundation for the divisions within Ulster and the cleavage between Ulster and the remainder of Ireland were laid.

In 1641, an estimated 30,000 Protestants were massacred in Ulster, but in 1649, the recalcitrant Irish were finally crushed by Cromwell and Catholic landowners were dispossessed in favour of Protestants. However, the final disruption between British and Irish Protestant and Catholic occurred in 1690 when James II, supported by Catholic Irish, was defeated by William of Orange at the Battle of the Boyne. The Protestant victory is still symbolical and is celebrated annually in Northern Ireland and parts of Britain. Following their defeat, a new penal code was introduced which denied to Catholics any right to citizenship or ownership of property, and the government of Ireland itself passed into the hands of a Protestant oligarchy. These events of 300 years ago still exercise a profound influence upon local, regional and national relationships today.

In the next century, following the American War of Independence, concessions were made and under the Catholic Relief Act, parliamentary franchise was extended to Catholics. However, two years later, in 1795, the rift between the two confessional groups was reinforced by the foundation of the Orange Order, a Protestant self-defence organization. None the less, despite the entrenched positions on both sides, an Act of Union was brought into force in 1801. Under this, the Irish were to be represented in the British Parliament by 28 peers and four bishops (House of Lords) and by 100 members in the House of Commons. A rebellion against the Act quickly followed and in 1835, the first major Catholic–Protestant riots of the century occurred in Belfast. However, the most far-reaching event in the mid-19th century was not political but human and economic. As a result of the failure of the potato crop in 1846, and constraints upon the entry of Irish wheat into the world market as a result of United States and British trading measures, there was a severe famine in which one million people (one-eighth of the Irish population) died. A further 1,250,000 emigrated and the face of Ireland was changed for ever. Unrest continued in Ireland and Irish Home Rule bills were introduced to Parliament in 1886, 1893 and 1913. On the last occasion, the bill received Royal assent, but implementation was suspended as a result of World War I.

Meanwhile, despite various bridge-building attempts, Protestant domination was strengthened in Northern Ireland. Furthermore, as a result of its commercial ties with Britain, the North developed a strong industrial base and in consequence, the gulf between it and the remainder of Ireland widened yet further. More importantly, the Protestants became more opposed to Home Rule, fearing that it would affect links with Britain and lead to economic decline.

The North–South division was also exacerbated by a number of other events. In 1898, the Ulster Unionist Party was established to provide a voice for Protestants supporting the Union. The Ulster Unionist Council, a provisional government, came into being in 1905 and in 1912, the Ulster Covenant was signed in Belfast by almost half-a-million people who pledged opposition to Home Rule. In the same year, the Ulster Volunteer Force was established as a Protestant defence militia. In the South, these developments were matched by the foundation of the nationalist Sinn Fein party in 1905 and, in 1913, by the establishment of the paramilitary National Volunteer Force. There followed a period of rebellion and great bitterness.

In 1916, the Easter Rising in Dublin was put down and in 1919, Sinn Fein members elected to the British Parliament organized their own parliament and declared Irish independence. Upon the suppression of the Sinn Fein movement, a guerrilla war broke out, leading to the signing of the Anglo–Irish Treaty on 6 December 1921. This established an Irish Free State, but Northern Ireland exercised its right to opt out and the Protestant Unionists retained control of six counties of Ulster. However, the Treaty divided Sinn Fein into those who saw it as a first step towards independence and the majority, led by De Valera, which rejected it on the grounds that it abandoned the principle of an all-Ireland republic. The result was a civil war, during which the Nationalist Volunteer Force adopted the name of the Irish Republican Army (IRA). In 1923, the pro-Treaty side prevailed, but opposition continued from the IRA and in 1925 the boundary between the Irish Free State and Northern Ireland was delimited.

Nevertheless, the underlying tensions remained and in December 1937, the independent state of Eire was proclaimed. The state constitution defines the 'national territory' as 'the whole island of Ireland'. It also accepts that, 'pending re-integration of the national territory', the constitution only applies in the 26 counties. As union seemed increasingly less likely, in 1939 the IRA launched a bombing campaign in mainland Britain.

After World War II, during which Eire

remained neutral, the situation became even more rigid when constitutional links between Eire and the United Kingdom were severed. The Republic of Ireland Act (1948) came into effect on Easter Sunday 1949, while at the same time the position of Northern Ireland as an integral part of the United Kingdom was guaranteed under the British government's Ireland Act (1949). The stage was thus set for the modern period.

The IRA re-emerged in the North in 1956 and from then on tension mounted until continuing violence eventually broke out in 1969. In 1970, the British government took over responsibility for security in the province and in the same year the IRA split into provisional and official factions. Violence reached a crescendo, following the shooting of demonstrators by British troops in Londonderry on 30 January 1972 and direct rule was imposed from London. Initial attempts at establishing a power-sharing Assembly, most notably as a result of the Sunningdale Agreement, were unsuccessful as were policies of internment and increases in police power under the Prevention of Terrorism Act. Violence in Northern Ireland and, to an extent in Great Britain itself, continued throughout the 1970s and 1980s. The Women's Peace Movement of 1975 had only a limited and short-term effect, and the focus of attention was directed more to political assassinations such as that of Lord Mountbatten in 1979 and IRA hunger strikers (1980–1). In 1984, the British Prime Minister, Margaret Thatcher, and her cabinet narrowly escaped death from an IRA bomb and in 1988 a furore was caused by the killing of three unarmed IRA terrorists by a Special Air Services (SAS) squad in Gibraltar (6 March 1988). Meanwhile, convictions resulting from bomb atrocities in a number of British cities have been demonstrated to be unsound.

In the continuing turmoil, one ray of hope was the Anglo–Irish Agreement, signed in November 1985. While confirming Northern Ireland's position with regard to the United Kingdom, it allowed an official consultative role to Dublin in the affairs of the province. However, even this was opposed by the more extreme and vociferous Protestant politicians in Northern Ireland. In 1990, a British–Irish consultative body met, but was boycotted by Unionists who refused to take part until the Anglo–Irish Agreement had been suspended. However in July 1992 British and Irish politicians (including the Unionists) met in face-to-face talks for the first time in decades.

The importance of the dispute is that it concerns the United Kingdom's only land border and therefore its only border with a fellow member of the European Community. Furthermore, it is the only border problem in Western Europe where positions are so entrenched that there has been military action over a period of some 70 years. Given the deep divisions within the society of Northern Ireland, the issue would seem to rank alongside that of the Arab–Israel dispute and possibly South Africa as one of the most intractable global problems.

Additionally, the continuing conflict has brought tragedy to many communities and has distorted British and Irish relations, both with each other and world-wide. Official statistics for the period 1971–88 show that, as a direct result of the problem 2,573 people were killed and almost 30,000 were injured in Northern Ireland.

Status

Despite several setbacks, the Anglo–Irish Agreement (1985) remains in place and offers a potential route towards settlement. However, the new Irish leadership is beset with domestic problems and the IRA campaign continues in Northern Ireland and on the mainland. Following the re-election of a Conservative government on 9 April 1992, a continuity in the British effort might have been expected. Nevertheless, the introduction of a new team at the Northern Ireland Office seems to indicate a growing preference for a military/security solution. There has even been a suggestion that internment could be re-introduced. The Loyalist Ulster Defence Association (UDA) was banned on 10 August 1992. The Republican Sinn Fein, unbanned in 1974, was not re-banned. Since neither internment nor the banning of organizations has had any noticeable effect on the situation in the past, this would seem to be a retrograde step. It would also be a clear illustration of how the long-running problem of Northern Ireland has distorted British policy-making. Like the Arab–Israel confrontation, the Northern Ireland issue has both internal and international aspects and therefore, Northern Ireland, like Israel and the Occupied Territories, is likely to remain, for the foreseeable future, a key global flashpoint.

References

Boyd, A. (1991), *An Atlas of World Affairs*, ninth edition, Routledge, London.
The Guardian (1992) 12 May.
Munroe, D. and Day, A.J. (1990), *A World Record of Major Conflict Areas*, Arnold, London.
Slowe, P. (1990), *Geography and Political Power*, Routledge, London.

50 Lake Nyasa (Malawi)

Description

A rift valley lake and the third largest in Africa, Lake Malawi (Lake Nyasa before independence) is situated at the southern end of the Great Rift Valley. It is 580 km long and varies in width from 24 to 80 km, with an area of 28,500 km². Nyasa is sometimes known as the 'calendar' lake as it is 365 miles long and 52 miles wide at its widest. Its surface is at an altitude of 437 m and it is almost 800 m deep. There are seasonal variations in lake level of up to 2 m and periodic variations on an 11-year cycle.

The Malawi–Mozambique border is approximately 1,560 km long, 328 km of which traverse the Lake. It was demarcated by the Anglo–Portuguese Agreement of 18 November 1954 which shifted the boundary from the eastern shore to the median line. Mozambique thereby obtained some 2,470 nml² of water surface.

The Malawi–Tanzania border is approximately 472 km long, 320 km of which follow the eastern shore of Lake Nyasa, down to latitude 11°34′ south, south of which the Mozambique border follows the median line.

History and importance

The original boundary line was based on 19th-century Anglo–German spheres of influence. An 1890 Anglo–German Agreement described the border as the eastern shore of Lake Nyasa. In practice, German sovereignty extended to the median line until 1922 when Tanganyika came under British control. Official British sources for the period 1916–34 showed the border as a median line, whereas those from 1947 to 1961 have generally abandoned the median line for an eastern shore border, in accordance with the 1890 Agreement.

When the Central African Federation of Rhodesia and Nyasaland was formed in 1953, this position was reaffirmed. However, Tanganyika, from its independence as Tanzania in 1961, rejected the change, on the grounds that it was arbitrary and illegal, since Tanzania was then a United Nations Trust Territory. In January 1967, Tanzania raised the boundary question again, restating its claims; Malawi replied that the matter would be given consideration.

Nevertheless, in September 1968, Malawi's President Hastings Banda declared that he considered that 'the real boundaries [of Malawi] are 160 km north of the Songwe river [in southwest Tanzania]: to the south, it is the Zambezi River (in Mozambique); to the west, it is the Luangwa River (in Zambia) and to the east it is the Indian Ocean.' These extravagant claims were based on Portuguese maps of the ancient empire of Maravi. On 27 September 1968, at Mbeya, Presidents Nyerere (Tanzania) and Kaunda (Zambia) took a common stand on these claims, which the Tanzanian government later dismissed as having 'absolutely no substance at all'. In 1977, President Banda was persuaded to drop the claims.

Status

Since 1967, there have been no further direct exchanges on the border issue and the situation must be considered dormant. However, the Malawi–Tanzania border must still be considered in dispute and, at times of discord, the issue is likely to be raised again, especially if navigation, fishing or potential mineral rights become contentious.

References

Brownlie, I. (1979), *African Boundaries: A Legal and Diplomatic Encyclopaedia*, Royal Institute of International Affairs, London.

Chambers World Gazetteer, (1988), Cambridge University Press, Cambridge.

Day, A.J. (ed.) (1984), *Border and Territorial Disputes*, Longman, London.

150 *Lake Nyasa (Malawi)*

Downing, D. (1980), *An Atlas of Territorial and Border Disputes*, New English Library, London.

United States Department of State (1964), *Malawi – Tanganyika and Zanzibar Boundary*, International Boundary Study No. 37, Office of the Geographer, Bureau of Intelligence and Research, Washington DC, 26 October.

United States Department of State (1971), *Malawi – Mozambique Boundary*, International Boundary Study No. 112, Office of the Geographer, Bureau of Intelligence and Research, Washington DC, 13 August.

51 The Oder-Neisse Line

Description

The Oder-Neisse Line became perhaps the most celebrated international boundary in the world when it was adopted as the border between East Germany and Poland. It was described in the Potsdam Agreement of 17 July to 2 August 1945: 'Pending the final determination of Poland's western frontier, the former German territories, east of a line running from the Baltic Sea, immediately west of Swinemunde, and thence along the Oder River to the confluence of the western Neisse River and along the western Neisse to the Czechoslovak frontier, including that portion of East Prussia not placed under the administration of the USSR, and including the area of the former Free City of Danzig, shall be under the administration of the Polish state.'

History and importance

In return for an eastern frontier on the Curzon Line, it had already been agreed in principle at Yalta (4-10 February 1945) that there should be a transfer of German territory to Poland. A secret agreement between the Provisional Polish Government and the Soviet Union, made on 21 April 1945, specified that the territories east of the Oder-Neisse Line should be placed under the Polish administration. The Western powers were not informed of this agreement for at least another two months.

The Polish position has always been that its claim was based on the Potsdam decision, which referred specifically to the western frontier of Poland and not a provisional line, and included such terms as 'former' German territories and 'administration' rather than 'occupation'. What is more, the decision resulted in the transfer of some 3.5 million people from Pomerania and Silesia, the two former German areas, incorporated into Poland.

At the Treaty of Zgorzelec (Gorlitz), on 6 July 1950, between the German Democratic Republic and Poland, the Oder-Neisse Line was accepted as a 'frontier of peace and friendship'. The final delimitation was agreed on 27 January 1951. Other treaties accepting the Line were those between the Federal Republic of Germany and the Soviet Union (12 August 1970) and between the Federal Republic of Germany and Poland (20 November 1970). The boundary was eventually reaffirmed by the 1975 Helsinki Final Act.

Status

After the unification of Germany, it was thought possible that claims might be made on the former German territories. However, at the Moscow Agreement of 12 September 1990, the Oder-Neisse Line was reaffirmed. Nevertheless, given the massive upheavals in Eastern Europe, this reaffirmation cannot be accepted as necessarily the final decision. Since the boundary effectively marks the current edge of NATO and the eastward limit of the EC, it is of more than national significance. As living standards in Western Europe rise and those in the East and the former Soviet Union decline, this border could become a flashpoint for confrontation with economic refugees.

References

'Germany – Poland Border Question Resolved' in *Boundary Bulletin*, No. 1, International Boundaries Research Unit, Durham University, p. 27.

Day, A.J. (ed.) (1984), *Border and Territorial Disputes*, Longman, London.

Downing, D. (1980), *An Atlas of Territorial and Border Disputes*, New English Library, London.

52 The Ogaden

Description

The Ogaden is an ill-defined area of semi-arid and arid land, lying between Ethiopia proper and Somalia. It is inhabited by tribes who are nomadic, pastoral and ethnically Somali. Estimates of the population vary from one to three million. Somalia itself comprises the former British Somaliland, an area of 174,000 km^2, with a population of 650,000 and the former Italian Somaliland, which covers 456,000 km^2 and has a population of 1.2 million. However, ethnic Somalis inhabit a further 328,000 km^2, beyond the present borders of Somaliland in Ethiopia, Djibouti and Kenya. Thus, in no way do political and ethnic boundaries accord. Indeed, for these largely nomadic peoples, political boundaries are of little relevance.

History and importance

On becoming independent in 1960, Somalia immediately abrogated the border treaty, agreed between Ethiopia and Britain in 1897. The aim appeared to be to inflame Somali irredentism and in 1963–4, there was a brief direct conflict between Somali and Ethiopian government troops over Somali support for guerrillas in the Ogaden. The Organization of African Unity (OAU) immediately took action to restore peace. Clashes between Kenya and Ethiopia ended with an agreement in 1968, but the continuing conflict with Ethiopia led Somalia to sign a pact with the Soviet Union which permitted the establishment of Soviet bases.

In 1974, the Emperor of Ethiopia, Haile Selassi, was overthrown and Somalia re-opened the border question. However, in 1977, the Somalis in Ogaden rebelled and received support from Somalian troops. Nevertheless, the full-scale war was won by Ethiopia, as a result of support from the Soviet Union, which flew in 20,000 Cuban troops. From then on, the Soviet Union continued to support Ethiopian troops in Eritrea, Tigre and on the Somalia frontier. In consequence, Somalia switched from Russian to United States protection. After the crushing defeat of Somali forces by Ethiopian forces at Jijinga in March 1978, the conflict subsided, but guerrilla forces remained active. By 1981, there were an estimated 1.5 million refugees in United Nations High Commission for Refugees (UNHCR) camps along the border between Ethiopia and Somalia.

The key importance of the region relates to its position in the Horn of Africa and the close proximity to Bab el Mandeb and the entry to the Red Sea. Before the break-up of the Soviet Union, there was effectively East–West confrontation on both sides of the strait. Ethiopia and South Yemen were both supported by the Soviet Union, while Somalia and to a certain extent North Yemen were supported by the United States. In the middle, Djibouti, independent since 1977, retained an uneasy relationship with France, which provided protection.

Status

The major problem of Somali irredentism remains unsolved and border clashes continue to be an ever-present possibility. President Barre, who came to power in 1969, was ousted in 1991, and the former territory of British Somaliland (originally incorporated into Somalia on independence) declared itself independent in the same year. Thus, at the present time, Somalia effectively lacks all government. At the same time, the government in Addis Ababba exerts only limited control over Ethiopia and little over its peripheral areas, while Eritrea is seeking independence. As a result, refugees have generally abandoned the camps and returned to the Ogaden. Thus, the Horn of Africa remains a major geopolitical flashpoint.

References

Boundary Bulletin, No. 1, (1991), International Boundaries Resesarch Unit, Durham University.

Brownlie, I. (1979), *African Boundaries: A Legal and Diplomatic Encyclopaedia*, Royal Institute of International Affairs, London.

Day, A.J. (ed.) (1984), *Border and Territorial Disputes*, Longman, London.

Downing, D. (1980), *An Atlas of Territorial and Border Disputes*, New English Library, London.

Griffiths, I.L.-L. (1985), *An Atlas of African Affairs,* Methuen, London.

Rais, R.B. (1986), *The Indian Ocean and the Superpowers*, Croom Helm, London.

53 The Panama Canal

Description

Apart from Suez, Panama is the most important canal in the world, strategically and economically. It is the only inter-oceanic waterway between the North West Passage, north of Canada and the Magellan Strait in southern Chile. It allows transit between the Atlantic and Pacific Oceans at a latitude not too far south of the world's most developed regions. Panama cuts through Central America at its narrowest point, the Isthmus of Panama.

The Panama Canal is 80.5 km in length and transit takes from eight to ten hours. The maximum permissible draft is 11.4 m to 12.2 m (depending on the lake water levels), and special Panamax ships, with maximum dimensions of 274.3 m by 32.3 m, have been designed to make maximum use of the Canal. A boundary delimited on 8 km either side of the Canal defines the 'Canal Area' (formerly known as the Canal Zone) and includes the cities of Cristobal and Balboa.

History and importance

In 1880, France made an initial attempt to breach the Isthmus of Panama using a design by Ferdinand de Lesseps, the designer of the Suez Canal. This was for a sea-level waterway, but after a few years it was abandoned, as too expensive. In 1903, the United States proposed the construction of a canal. When Colombia was slow in its response, the United States organized the secession of Panama as an independent state but under American tutelage. On 18 November 1903, the United States then concluded a treaty with the newly independent republic, giving the United States unilateral control of canal operations, together with the civil and military administration of the newly delimited Canal Zone.

During the period 1904 to 1914, a lock-based canal was constructed by the United States and officially opened on 15 August 1914. From 1975, a widening programme was started, and in 1977 a new Panama Canal Treaty was negotiated. This took effect on 1 October 1979 when the Canal Zone (Canal Area) was formally transferred to Panamanian sovereignty, along with the cities of Cristobal and Balboa, the dry docks, the trans-isthmus railway and the naval base at Coco Solo. According to the treaty, by the year 2000, control of the Canal passes to Panama and the United States must have removed all its bases from the country.

Panama is vital to American strategic interests, particularly for the redeployment of its navy, between the Atlantic and the Pacific. 'Sea Lift', the replenishment of Europe from the United States, is dependent upon the rapid transit of vessels between the oceans. Thus, in the year 2000, the United States will lose control of a key element in its global reach. The Canal is also vital to world shipping in general and East–West trade in particular. For instance, the Canal reduces the passage from London to Auckland, New Zealand, to 11,380 nml, as opposed to 12,670 nml via the Suez Canal or 12,480 nml by the Cape of Good Hope. It is also a key part of the US coastal shipping infrastructure, particularly for bulk commodities (the passage between San Francisco and New York via the Panama Canal is 5,263 nml, as opposed to 13,100 nml via Cape Horn).

Status

Future problems include the political uncertainty associated with the American departure from Panama and increased congestion. A recent poll (1991) showed that 65 per cent of Panamanians would prefer continued American military presence and 62 per cent have no confidence in Panama's ability to operate the waterway on its own. They are worried further that without investment, the Canal might become obsolete. However, it must be remembered that such worries were expressed

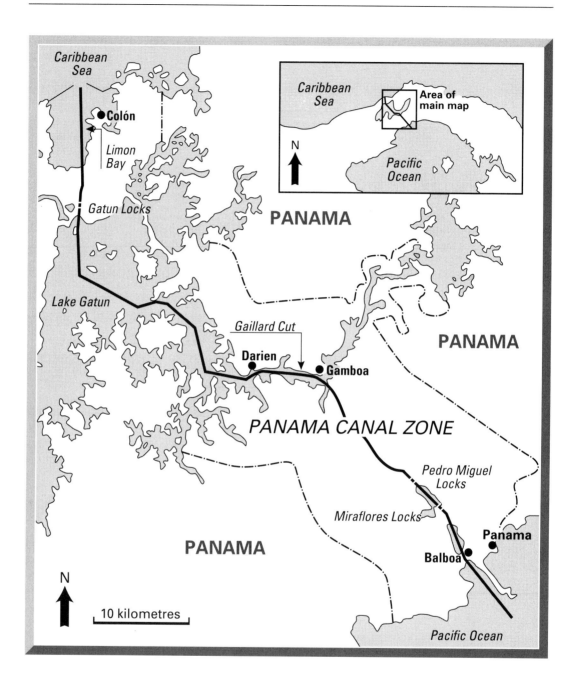

about Egypt and the Suez Canal in 1956 at the time of the 'Suez Crisis' and these have since proved totally unfounded.

Use of the Canal is still rising, but at rates far lower than the 8–10 per cent per annum rises of the early 1980s. A Commission established by the governments of the United States, Japan and Panama found that of 160 cargo vessels under construction (April 1991), 156 could pass through the locks of the Canal and 60 had been specifically designed as Panamax ships. Thus, continued traffic for at least the next 20 years seems assured.

However, there remains the problem of congestion, in that only one ship at a time can pass through the Gaillard Cut in the central highlands. Railways might offer an alternative land bridge across the United States and an oil pipeline across Panama itself was completed in 1982. To overcome the problems of the Canal, various alternatives have been put forward:

(a) widening the cut at a cost of about $400 million;
(b) constructing a new set of larger locks at a cost of $3–4 billion;
(c) building a sea-level canal at a cost of $8–15 billion;
(d) developing a fast-transit rail link; or
(e) constructing an alternative canal–rail link through Nicaragua.

While the Panama Canal has never been a global flashpoint, there clearly is potential for geopolitical problems, particularly after the year 2000, if the Canal were to be deliberately blocked. Not only would there be major strategic and economic implications for international trade, but also a major strategic impasse for trans-oceanic transfer of naval forces. Even if land-based alternatives could be used for freight transhipment as a temporary or semi-permanent substitute, there would remain a major problem if the United States had quickly to redeploy the Pacific Fleet to the Atlantic or vice-versa.

References

Boyd, A. (1991), *An Atlas of World Affairs*, Routledge, London.
The Economist Atlas (1989), Economist Books, Hutchinson, London.
The Economist (1991), 6 April.
Times Atlas of Oceans (1983), Times Books, London.

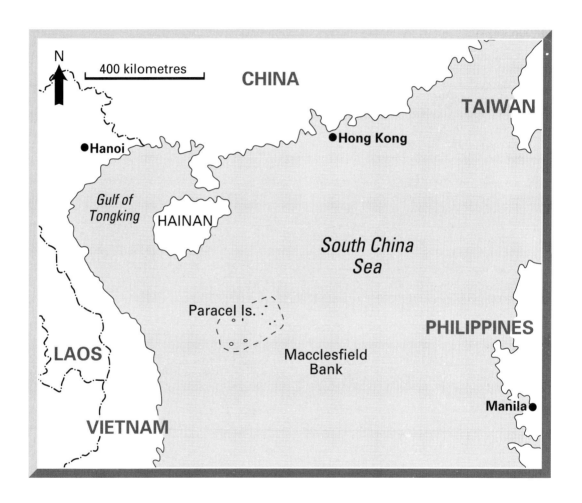

54 The Paracel Islands

Description

The Paracels, known as the Xisha Qundao in Mandarin Chinese and the Quan Doa Hoang Sa in Vietnamese, lie approximately equidistant some 150-200 nml from Hainan Island and Vietnam. Two clusters are recognized, the Amphitrite and the Crescent groups. There are approximately 15 islets and a large number of reefs and shoals, scattered in a roughly oval shape, approximately 160 nml long in the Gulf of Tonkin. Five islands are occupied by troops from China.

As a result of their geographical proximity to China and Vietnam, both nations claim the Paracels. The Chinese base their case on the fact that the islands were discovered by Chinese navigators and used by Chinese fishermen for centuries, before being brought under Chinese administration from the 15th century. None of this was disputed until the 1930s. Vietnam, on the other hand, claims that the Paracels form part of the 19th-century empire of Annam, and before then, no country was in control of them.

History and importance

In 1932, the Paracels were annexed by France, which had colonized Indochina, and in 1939 they were occupied and taken over by Japan. At the San Francisco Treaty in 1951, Japan renounced its claim, but no statement was made as to the ownership of the islands. On 16 January 1974, in the 'One Hundred Minute War', the Chinese drove South Vietnamese forces out of the Paracels and established a garrison which has remained there ever since.

The islands are strategically important as the main Singapore-Hong Kong shipping route passes between them and the Macclesfield Bank to the east. There is also a very strong potential for seabed hydrocarbon exploitation.

Status

Ownership of the Paracel Islands is still an issue of dispute between China and Vietnam. At present, China occupies the islands and, given its military strength compared with that of Vietnam, little change seems likely. However, as the Americans and Russians withdraw from the area, and regional powers become established in the Pacific Rim, conflict must remain a possibility.

References

Boundary Bulletin, No. 1, (1991), International Boundaries Research Unit, Durham University.
Boundary Bulletin, No. 2 (1991), International Boundaries Research Unit, Durham University.
Day, A.J. (ed.) (1984), *Border and Territorial Disputes*, Longman, London.
Dzurek, D.J. (1985), 'Boundary and Resource Disputes in the South China Sea' in E.M. Borgese and N. Ginsburg (eds), *Ocean Yearbook 5*, University of Chicago Press, pp. 254-84.
The Economist, (1989), 21 May.
Park, C. (1980), 'Offshore Oil Development in the China Seas: Some Legal and Territorial Issues' in E.M. Borgese and N. Ginsburg (eds), *Ocean Yearbook 2*, University of Chicago Press, Chicago, pp. 302-16.
Prescott, J.R.V. (1985), *The Maritime Political Boundaries of the World*, Methuen, London.
Weatherbee, D.E. (1987), 'The South China Sea: From Zone of Conflict to Zone of Peace?' in L.E. Grinter and Y.W. Kihl (eds), *East Asian Conflict Zones*, Macmilan, London, pp. 123-48.

55 The Rann of Kutch

Description

A desolate area, some 20,500 km² in extent, the Rann of Kutch comprises salt marshes, brackish lakes and isolated rocky elevations on the Indo–Pakistani border. Between June and November, it is largely inundated, forming a shallow body of salt water, with a maximum depth of between one and two m. As the southwest monsoon season ends, the water level drops rapidly and the area returns to predominantly salt flats. The northern edge of the Rann of Kutch provides the boundary between India (Gujarat) and Pakistan (Sind), awarded by the Indo–Pakistan Western Boundary Case Tribunal. This stretch of boundary is some 403 km long.

History and importance

Prior to independence, the British Indian Province of Sind and the British suzerainties of Kutch, Santalpur, Tharad, Suigam, Wav and Jodphur abutted on to the Rann of Kutch. With the India Independence Act of 18 July 1947, Sind was allotted to Pakistan, while on the expiry of the suzerainties, the remainder acceded to union with India. Previously, following a dispute over the southern limits of Sind in 1913, a partial demarcation of the boundary was achieved in 1923–4.

In July 1948, Pakistan raised the question of the Sind–Kutch frontier delimitation, east of this demarcated boundary. Although it arose periodically, the issue remained low key until 1965, when there were frontier clashes. Hostilities terminated when Britain persuaded the combatants to submit the dispute to the Indo–Pakistan Western Boundary Case Tribunal. In February 1968, the Tribunal presented its award, determining the boundary. The award referred to approximately 9,000 km² of uninhabited territory.

The case rested primarily on a definition of the Rann of Kutch. Pakistan submitted evidence to prove that Sind extended south of the Rann before and after the period of British administration. Furthermore, the Rann was a marine feature and therefore the equitable distribution of land depended upon a median line. In other words, the Rann itself was treated as a wide boundary area. The equidistant line claimed by Pakistan roughly accorded with latitude 24°N. In contrast, India contended that the boundary ran along the north of the Rann and cited the 1923–4 demarcation line in support of its case. By a majority, the Court established a compromise, close to the northern shore of the Rann of Kutch.

The importance of the area depends partly upon the potential for oil discovery. It is also significant in terms of the continuing conflict between India and Pakistan in which neither wishes to lose face.

Status

In comparison with the conflicts and potential problems along other parts of the India–Pakistan border, the Rann of Kutch dispute can be considered dormant. However, if significant hydrocarbon resources are discovered, the issue could be raised again. Furthermore, should the relations between India and Pakistan deteriorate into conflict, for example, over Kashmir, fighting could again take place over the salty wastes of the Rann.

References

Boyd, A. (1991), *An Atlas of World Affairs*, Routledge, London.
United States Department of State (1968), *India – Pakistan (Rann of Kutch)*, International Boundary Study No. 86, Office of the Geographer, Bureau of Intelligence and Research, Washington DC, 2 February.

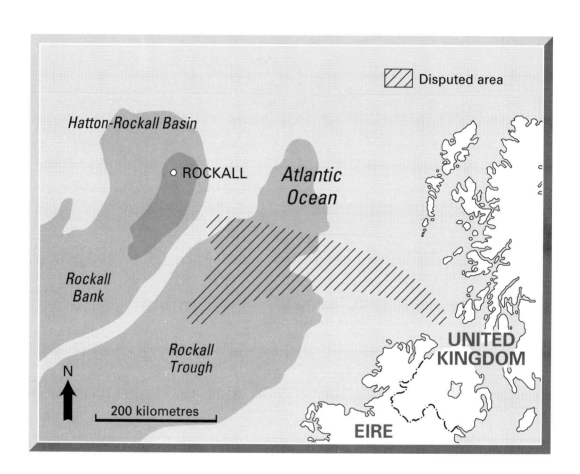

56 Rockall

Description

Located on the Rockall Bank, some 225 nml west of the Hebrides, Rockall is an extremely isolated islet. Its exact location is 57°36′N, 13°41′W. The islet, consisting of a very small rock platform, is uninhabitable but has been formally annexed by the United Kingdom. Apart from the United Kingdom, claimants to Rockall are Ireland and, from 1985, Iceland and Denmark.

The dispute between the United Kingdom and Ireland rests on the underlying legal principle for delimitation. According to Britain, there should be a strict median line, using all offshore islands as base points. In contrast, the Irish approach is for an 'equitable' equidistance line, based on such factors as habitability and population. The claims of Iceland and Denmark (through the Faeroes) use the argument that the Rockall–Faeroes plateau is a natural prolongation of their landmass, not that of Britain or Ireland. The Rockall Deep in fact separates the continental shelf off the British Isles from a series of banks which link Rockall with the Faeroes and Iceland.

History and importance

During the 1960s and 1970s, there were unsuccessful negotiations between Britain and Ireland, and it was not until 1974 that Ireland proposed the equidistance line. Britain reacted to this by designating as British 52,000 nml^2 of sea, 300 nml west of Scotland, including areas on the Irish side of the proposed equidistance line. The Irish response was to designate some areas across the proposed British median line, closer to Rockall than Ireland as Irish. In 1976 and 1977, there followed a further round of designations in the opponent's claimed areas and in February 1977 the Irish repeated their proposal, originally made in April 1976, for the dispute to go to arbitration. On 21 February 1977, Britain agreed to arbitration.

Eventually, in 1980, Ireland agreed to a five-man tribunal, but progress was minimal as both sides waited for results from several other International Court of Justice (ICJ) cases. In 1985, the emergence of the claims by Iceland and Denmark spurred Britain and Ireland to move away from arbitration to negotiated practical settlement, as it was considered that an arbitration court would need to take the new Scandinavian claims into account.

As a result, on 8 November 1988, agreement on the delimitation of the continental shelf was reached and signed by the foreign ministers of Britain and Ireland. The result was unique in such settlements, in that it comprised a stepped line. Such a line might be useful in the delimitation of oil exploration concessions, but it poses obvious problems for ships trying to locate their exact position. In the Treaty, no mention was made of Rockall.

The importance of Rockall is that it provides the basis for claims to huge areas of sea and sea-bed. A claim for as much as 120,000 nml^2 of sea space would not be unreasonable and this is clearly vital in an area where there is a high potential for hydrocarbon discoveries.

Status

The agreement reached in 1988 effectively preserves the status quo. However, the question of the ownership of Rockall is still unsettled, according to Irish officials. Despite the heroic efforts of a Royal Marine, who spent some days on Rockall as a symbolic gesture, the islet is uninhabitable and therefore should not provide the basis for offshore claims.

It appears now that Rockall will be treated as a separate issue. Ireland has waived sea-bed rights around the rock, except for the 12-nml limit which remains in dispute. Nevertheless, the Scandinavian claims are still outstanding and could cause problems in the future. Furthermore, developments in Northern Ireland might also cloud the issue.

References

Boyd, A. (1991), *An Atlas of World Affairs*, Routledge, London.

Lysaght, C. (1990), 'The Agreement on the Delimitation of the Continental Shelf between Ireland and the United Kingdom', *Irish Studies in International Affairs*, 3, No. 2, pp. 83–109.

Symmons, C.R. (1989), 'The U.K./Ireland Continental Shelf Agreement 1988: A Model For Compromise in Maritime Delimitation', in *International Boundaries and Boundary Conflict Resolution*, 1989 Conference Proceedings, International Boundaries Research Unit, University of Durham, pp. 387–412.

57 The Sahel

Description

The Sahel comprises a vast semi-desert swathe of western Africa, along the southern edge of the Sahara Desert. It is characterized by low and unreliable rainfall and has become the focus of world attention as a result of mass starvation-related drought. The main Sahelian countries are Mauritania, with an area of over one million km^2 (40 per cent of which is Saharan desert and 30 per cent semi-desert) and a population of 2 million, giving it a density of 2 per km^2. Mali is 1.24 million km^2 in area and has a population of 7.6 million, with a density of 6.1 per km^2. Niger covers 1.26 million km^2 is almost two-thirds desert. Its population is 6.1 million, with a density of 5 per km^2. Senegal and Burkina Faso are considerably smaller and have significantly higher population densities. Senegal covers 196,192 km^2 and has a population of 7 million, with a density of 35.5 per km^2. The area of Burkina Faso is 274,122 km^2 and its population is 7.9 million, with a density of 29.1 per km^2. These are all extremely poor countries, with basic pastoral economies and a lack of resources, particularly water.

History and importance

Media attention was first brought to the Sahel in 1973 when over 100,000 people died as a result of widespread famine. Throughout the region rainfall is low, less than 600 mm per annum, and falls in a single short rainy season. Moreover, it is highly unpredictable in both its total and occurrence. Furthermore, the high temperatures prevailing result in high rates of evaporation and evapotranspiration. Therefore, the overall effectiveness of even the minimum fall is reduced.

There has been a good deal of debate over whether the very low rainfall received in certain years was a departure from the normal or merely part of a cycle. No convincing evidence has been produced to show that there is a trend towards increasing dryness and indeed it would be difficult to do this, given the scarcity of data.

Furthermore, drought may be defined in a number of ways and the effects of an increasing population and over-use of the land may be more important than any climatic factors. However, there do appear to be long-term cyclical variations in rainfall and during the 20th century, the Sahel has suffered severe droughts in what is approximately a 30-year cycle. The median years of dry periods were 1913, 1942 and 1971, when, respectively 59 per cent, 79 per cent and 70 per cent of 'normal' rainfall was experienced. In each case, drought conditions prevailed for several consecutive years as, for example, in the period 1968–72, when rainfall averaged only 81 per cent of the 'normal'.

Conditions were made worse by the fact that the 1968–72 period was preceded by above-average rainfall in the 1950s and early 1960s. As a result, there was an explosion in the population and number of livestock and ever more marginal areas were exploited, leading to increasing desertification. In the Sahel, births averaged 45–50 per 1,000 and deaths, owing to improving health care, registered 20–25 per 1,000. Thus, the meagre and sparse resources of the Sahel have increasingly to support greater population.

The importance to the West of the Sahel and its problems is purely humanitarian, since there are no overriding economic or political concerns. Thus, there is invariably a lag in recognition and response times, to any crisis. The problem is compounded by difficulties of access and the low densities of population which militate against the distribution of aid. The few key distribution points rapidly become vast refugee camps, with populations completely out of proportion to anything the local area might be able to support.

Logistics are another crucial problem, with a severe lack of railways and tarred roads. From Accra (Ghana) to Ouagadougou (Burkina Faso) is

168 The Sahel

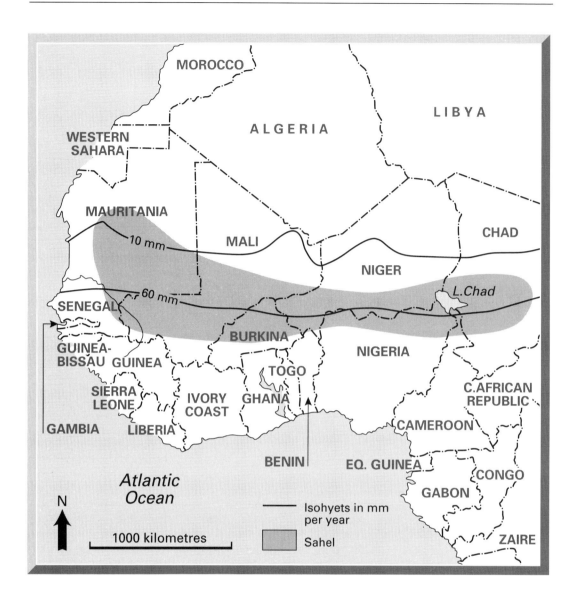

1,000 km, and from Dakar (Senegal) to Bamako (Mali) is more than 1,300 km. It is 1,200 km from Bamako to Gao (Mali) and 1,500 km from Niamey (Niger) to Lake Chad. In the whole region, there are only two modern airports, those at Bamako and Kano (Nigeria).

Status

Already drought-stricken, impoverished and debt-ridden, the countries of the Sahel are gradually being enmeshed in political turmoil. In particular, the nomadic Tuareg people moved north into Libya and Algeria in the 1970s and 1980s and some served in the Libyan Army. They are now returning to the Sahel and raiding has occurred in northern Mali and Niger, with conflict overspilling into Algeria and Mauritania. Demands for Tuareg autonomy remain unmet and guerrilla or civil war is likely to continue, even if a peace agreement should be reached, as the moderate Tuareg leadership has little control over many of the rebel gangs.

It is realized that many of the political problems would be resolved if the ecological deterioration could be controlled. There is a plethora of projects and schemes, ranging from tree-planting and terracing to small-scale irrigation, but the fact remains that in many parts of the region, the population is now too large to be supported by the local environment. The outlook for the Sahel is therefore gloomy, but, despite the influence of television, the effect of further catastrophes upon the West is likely to remain limited.

References

Boundary Bulletin, No. 1, (1991), International Boundaries Research Unit, Durham University.

Boundary Bulletin, No. 3, (1992), International Boundaries Research Unit, Durham University, January.

The Economist Atlas (1989), Economist Books, Hutchinson, London.

Griffiths, I.L.-L. (1985), *An Atlas of African Affairs*, Methuen, London.

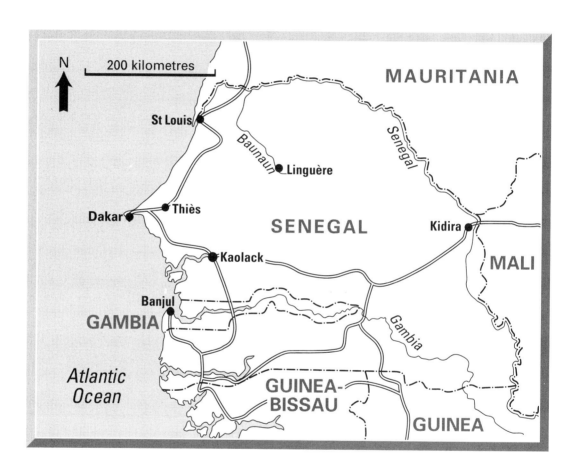

58 Senegambia

Description

Senegambia was a confederation of Senegal and Gambia, located on the far west coast of Africa, and dissolved in 1989. Gambia is a small enclave, effectively the banks of the Gambia River, completely surrounded, other than at its exit to the sea, by Senegal. Apart from microstates such as the Vatican, the only independent country in the world completely surrounded by another state is Lesotho. Gambia is thus very dependent upon its erstwhile confederal partner.

The area of Senegal is 196,192 km² and it has a population of 7 million, with a density of 35.5 per km². The population is 91 per cent Muslim and 6 per cent Christian. Senegal became independent in 1960 and since then has sustained a multi-party democracy. Gambia is on a totally different scale, having an area of 11,295 km², a population of 700,000 and a population density of 61 per km². It is 85 per cent Muslim and 3 per cent Christian. The country, which became independent in 1965, comprises largely mangrove swamps and is, on average, only 25 km wide along the Gambia River.

History and importance

James Island was occupied by British traders in 1618, in 1816 Bathurst was established and in 1821 these two trading posts became the colony of Gambia. Later, in 1823, MacCathy's Island and Albreda were added. Meanwhile, in 1854, having been originally based at St. Louis, on the Senegal River, the French established Dakar.

During the 1850s, and early 1860s, Gambia received French help in putting down tribal insurrection, and in 1876 the French proposed the incorporation of Gambia into the colony of Senegal. However, the proposals were refused by the British Parliament. During the 1880s, there was a great expansion of French trading activity in West Africa and this proved formidable competition for British interests. Between 1887 and 1888, the interior of Gambia was occupied, but on 10 August 1889, a treaty, delimiting the frontiers, was signed with the French. The delimitation was based on lines parallel to, but 10km from, the right and left banks of the Gambia River.

When, in the 1960s, both states became independent, both were stable and relations were good. By 1973, it was realized in Gambia that union with Senegal was an 'inevitable and necessary development' and from then on there was a succession of cooperative schemes, including the trans-Gambia highway and the formation of the Gambia River Development Organization. In July 1981, when the president of Gambia was on a visit to London, there was a coup, but this was defeated with help from Senegal and the British SAS. This merely accelerated talks of union and on 1 January 1982, the Confederation of Senegambia was formally declared.

In January 1983, the Confederal Cabinet was announced with a president from Senegal and a vice-president from Gambia. On 13 January, the Confederal Parliament met with 20 Gambian and 40 Senegalese members, at Dakar.

Senegambia had relatively little economic or political significance, but was symbolically important as a union of the former colonial systems of Britain and France.

Status

During seven years of union, progress towards integration was still extremely slow. The two states remained separate, with their respective British and French systems. The instability of the union and its collapse in 1989 illustrate the strong persistence of former colonial boundaries in Africa, although it seems likely that pressures for amalgamation will continue to be exerted in the future.

References

Downing, D. (1980), *An Atlas of Territorial and Border Disputes*, New English Library, London.
The Economist Atlas (1989), Economist Books, Hutchinson, London.
Griffiths, I.L.-L. (1985), *An Atlas of African Affairs*, Methuen, London.
Porter, A.N. (1991), *Atlas of British Overseas Expansion*, Routledge, London.

59 The Senkaku and Ryukyu Islands

Description

The Senkaku Islands, known to the Chinese as the Tiaoyu or Tiao Yu Tai Islands are situated some 200 nml west of Okinawa and about 100 nml northeast of Taiwan. They comprise a small, uninhabited group of some five coral islands and scattered islets. The largest is only 4 km long and 1.5 km wide. Although physically distinct, the Senkaku Islands are frequently considered in association with the Ryukyu Islands.

The Ryukyu Islands form a chain 650 nml long, dividing the East China Sea from the Pacific. They stretch in virtually a straight line, south-southwest from Kyushu (Japan) and have a total area of 2,246 km^2 and a population of just over 1.1 million (1980). The principal island is Okinawa, which is 1,176 km^2 in area and lies 528 km from Kyushu. Given their location, both groups of islands have economic and strategic significance.

History and importance

Formerly an independent kingdom, the Ryukyu Islands were first invaded by the Chinese in the 7th century and then, in the 17th century, by the Japanese. At that time, they paid tribute to both both countries. However, in 1874, China relinquished its claims to the islands and they became part of the Japanese empire. In 1895, the Japanese occupied the Senkaku Islands.

After World War II, during which Japan extended its control over most of Southeast Asia including island groups from the Gilberts to the Andamans, the Senkakus and Ryukyus were placed under American military control (1945). In 1951, at the San Francisco Peace Treaty, the Senkaku Islands were included with the Ryukyu group. In 1951, while still under American occupation, a native civilian governor was appointed and at the Treaty between Japan and the United States, finalized on 14 May 1972, the Ryukyus, including Okinawa and the Senkaku Islands, reverted to Japan. However, relations between Japan and the United States remained strained since, in the case of Okinawa, the United States retained rights to a base. Of even greater symbolic significance for the Japanese was the possible nuclear use of the base.

However, prior to this, on 11 June 1971 Taiwan claimed the Senkaku Islands and on 30 December in the same year, the People's Republic of China claimed the Senkakus as part of its greater claim on Taiwan. In February 1972, the government of Taiwan announced the incorporation of the Senkakus in Taiwan. On 17 February, Japan protested and later issued a document, stating that under the Treaty of Shimanoseki (1895) the islands had been incorporated into Japanese territory, along with Taiwan itself and the Pescadores Islands. This document further established that Chinese rule did not extend to the Senkakus.

On 12 August 1978, Japan and China signed a Treaty of Peace and Friendship, but in April of that year, Chinese fishing boats had begun operating in the territorial waters of the Senkakus. Later that month, the fleet was withdrawn and the vice chairman of the Standing Committee of the National People's Congress of China stated that all efforts would be made to avoid conflicts over the Senkaku Islands. In August 1978, following the Treaty, it was stated by the Japanese spokesman that China had, for all practical purposes, recognized Japanese control of the Senkaku group.

On 21 October 1990, Japanese coastguard patrol boats turned back two Taiwanese vessels, seeking to assert Taiwan's sovereignty over the Senkakus. The incident provoked anti-Japanese protests in Hong Kong.

The Senkaku Islands are of little intrinsic value, but sovereignty, together with a 200-nml EEZ claim, would offer potentially rich hydrocarbon exploitation and fishing rights. There is clearly some strategic importance in the position of the islands but far less than that enjoyed by the

Ryukyus and Okinawa, in particular, which guard the approaches to and exits from the Sea of Japan and Japan itself. The Ryukyus also, of course, offer large-scale sea-bed and fishing rights.

Status

Although there are still some outstanding claims by China and Taiwan, it seems unlikely that there will be actual conflict over Japanese occupation of the islands. However, the memory of Japanese militarism in the region remains influential and any sign of renewed Japanese assertiveness would produce a very delicate situation.

References

Boundary Bulletin, No. 1, (1991), International Boundaries Research Unit, Durham University.

Chambers World Gazetteer, (1988), Cambridge University Press, Cambridge.

Day, A.J. (ed.) (1984), *Border and Territorial Disputes*, Longman, London.

Downing, D. (1980), *An Atlas of Territorial and Border Disputes*, New English Library, London.

60 The Shatt al Arab

Description

The Shatt al Arab results from the confluence of the Tigris and Euphrates Rivers at Qurna. Approximately halfway along its course, a left bank tributary, the Karun, with its watershed entirely in Iran, greatly enhances flow. The total length of the waterway from its confluence to the Persian/Arab Gulf is 209 km and it has an average width of 400 m, reaching 1 km in the estuary. The average depth from Fao to Basrah is 7 m and the navigable channel, formerly some 3 m wide, has been extended by Iraqi dredging to over 7 m at low tide. For much of its length, the Shatt al Arab flows through low marshy ground and a major part of its discharge is absorbed by lakes and swamps.

The Shatt al Arab marks not only the border of Iraq and Iran, but also that of the Arab and Persian worlds. Throughout much of history, therefore, it has formed an important boundary between the people of the irrigated lowlands and those of the mountains.

History and importance

Before 1847, the boundary between the Ottoman and Persian Empires was vague and completely undelimited in the area of the Shatt al Arab. However, in 1847, the Treaty of Erzurum, which resulted from the work of a boundary commission, including Persian, Turkish, British and Russian officials, was signed (19–31 May 1847). This delimited the boundary on the left, east or Persian bank of the Shatt al Arab, leaving the waterway under Turkish sovereignty but allowed freedom of navigation.

The Constantinople Protocol of 4 November 1913, delimited the entire boundary in detail and stated specifically that the Shatt al Arab, except for certain islands, was under Turkish sovereignty. By October 1914, the demarcation of the boundary was complete. It followed the low-water mark on the Persian bank of the Shatt, with the exception of the area around Khorramshahr, where the line followed the thalweg (deepest part of the waterway).

In 1934, Iran (the name Iran was used for Persia from 1927) challenged the validity of the Treaty of Erzurum and the Constantinople Protocol. There followed, on 4 July 1937, a Frontier Treaty between Iraq and Iran which reaffirmed the boundary established by the Constantinople Protocol and the minutes of the 1914 Boundary Commission, with the exception of the Abadan area, where, as in the Khorramshahr area, the boundary was moved from the low-water mark to the thalweg.

This change removed one of Iran's major grievances, but it remained worried about freedom of navigation and pressed for a thalweg boundary throughout the Shatt al Arab. Finally, on 6 March 1975, in a Joint Communiqué, between Iran and Iraq, signed in Algiers, Iraq conceded a thalweg boundary throughout the length of the Shatt. In return, Iran ceased providing aid for Kurds in northern Iraq.

In 1980, Iraq abrogated the Algiers Communiqué and the two countries moved to war, much of which was fought in the vicinity of the Shatt al Arab. Indeed, one of the major events was the occupation of Fao on the right bank of the Shatt al Arab, by Iranian troops. In 1990, following the invasion of Kuwait, and the resultant United Nation's sanctions, Iraq looked to support from Iran. Iran provided the only boundary which could not be easily monitored by coalition forces and, although it received little additional support in return, Iraq agreed to return to the thalweg boundary.

While fishing and irrigation are of importance in the Shatt al Arab, the overwhelming significance is with regard to navigation. Apart from the port and naval base of Umm Qasr, which is adjacent to the Kuwait boundary and restricted by Kuwaiti territorial waters, the Shatt al Arab provides Iraq's only outlet to the sea. Thus, it is fundamental to Basra and the whole Tigris–Euphrates valley.

On the other hand, Abadan and Khorramshahr, Iran's most important ports until the Iran-Iraq War, also depend upon the Shatt for trade. The waterway was particularly critical for the export of oil by both countries, but it is now blocked by more than 80 vessels sunk during the Iran-Iraq War. As a result, Iran has improved facilities on its Gulf coast, but Iraq is effectively land-locked.

Status

The waterway remains closed and therefore Iraq has effectively no access to the Persian/Arabian Gulf. According to the 1990 agreement between Iraq and Iran, the boundary between Iran and Iraq runs along the thalweg. However, given the volatility of the region, and the continuing friction between Iran and Iraq, it seems most unlikely that the Shatt al Arab will not again be the scene of confrontation. Therefore, it remains one of the key geopolitical flashpoints.

References

Amin, S.H. (1984), *Political and Strategic Issues in the Gulf*, Royston, Glasgow.

Anderson, E.W. and Rashidian, K. (1991), *Iraq and the Continuing Middle East Crisis*, Pinter, London.

Blake, G.H., Dewdney, J. and Mitchell, J. (1987), *The Cambridge Atlas of the Middle East and North Africa*, Cambridge University Press, Cambridge.

61 The Sinai Peninsula and Taba

Description

The Sinai Peninsula is a rugged area of desert, extending from the head of the Red Sea to the Mediterranean and bounded by the Gulf of Suez and the Suez Canal to the west and the Gulf of Aqaba and Israel to the east. It effectively insulates Egypt from Israel and Saudi Arabia. The area of Sinai is 60,174 km^2 and it rises to heights of 2,637 m in Jebel Katherina and 2,286 m in Jebel Musa (Mount Sinai). The population is said to be just over 10,000.

History and importance

Traditionally, Sinai had been part of the Egyptian Empire, acting as a buffer province, but in 1906, it was decided by Britain, then effectively administering Egypt, to demarcate a boundary between Egypt and Palestine to secure the Suez route to India. On 10 January, there was confrontation over the idea and on 1 October, Britain imposed an administrative line which was later surveyed and demarcated in 1912–13. This line is basically the boundary line of today.

In 1915, Egypt came officially under British control but gained independence in 1922 as a monarchy. On 24 February 1949, the Armistice Agreement between Israel and Egypt formalized this division, the Negev going to Israel and Sinai to Egypt. Between then and October 1956, there were 11,650 incidents on the Israel–Egypt–Sinai border. One effect was that between 1951 and 1956, over 400 Israelis were killed as a result of infiltration from Sinai.

On 29 October 1956, in cooperation with Great Britain and France, Israel launched an offensive against Sinai, achieving large-scale rapid success. Almost immediately, the United States and the Soviet Union brought pressure to bear to halt the conflict (the so-called Suez Crisis) and on 6 November, Great Britain accepted the cease-fire. The Anglo–French troops rapidly withdrew and Israel evacuated Sinai, but retained the Gaza Strip and a coastal stretch from Eilat to Sharm el Sheikh (until March 1957). United Nations Emergency Force (UNEF) forces were then deployed in Sinai, but incidents continued along the 1949 Armistice Line.

On 5 June 1967, Israel made a series of preemptive strikes and, crucially, destroyed the Egyptian Air Force while it was still on the ground. The result of the 'Six-Day War' was a rapid and complete victory for Israel, which then occupied Sinai up to the Suez Canal. Although the war officially ended quickly, from then until 1970 Israel and Egypt continued a series of smaller conflicts along the Suez–Sinai front for over 1,000 days, but there were no territorial changes.

On 6 October 1973, the fourth Arab–Israel war, the Yom Kippur War, broke out. A surprise attack by Egypt pushed Israeli forces back and by the cease-fire of 24 October, Egypt had established itself along much of the east bank of the Suez, north of Ismailiya and along a coastal strip, some 5–16 km wide, down the western side of Sinai. However, in the southern sector, Israel had launched a successful counter-attack across the canal, to control some 1,600 km^2 of Egyptian territory west of the Great Bitter Lake. During the war 2,400 Israelies were killed and Arab losses are unknown but were considerably larger.

On 18 January 1974 and 4 September 1975, disengagement agreements were reached on the Sinai front. The first agreement established a buffer zone with UNEF contingents and all territory west of the canal, together with the east bank of the canal, passed to Egypt. Israel withdrew some 20 km east of the canal, but retained the rest of Sinai, including the strategic Mitla and Giddi Passes, the Bir Gafgafa defence zone behind them and Sharm el Sheikh, commanding the Strait of Tiran.

Following the second agreement, Israel withdrew a further 20–40 km to the eastern end of the Mitla and Giddi Passes and the vacated area

180 The Sinai Peninsula and Taba

became a new United Nations buffer zone. The former buffer zone was redesignated the new Egyptian Limited Forces Zone. The Israelis also vacated the coastal strip down the Gulf of Suez, together with the Abu Rudeis and Ras Sudar oilfields. As a result, Israeli ships gained the right of passage through the Suez Canal.

Two years later, on 19 and 20 November 1977, President Sadat of Egypt undertook the first official visit to Israel by any Arab leader, on the second day of his visit addressing the Knesset. The reciprocal visit took place on 25 and 26 December, when Prime Minister Begin was officially received for talks in Ismailiya. The longer-term result of these meetings was that on 26 March 1979, at Camp David, the American presidential retreat in Maryland, through the diplomatic efforts of President Carter, Israel and Egypt signed a Peace Treaty (the Camp David Accords), which was ratified on 25 April at the American surveillance post at Um-Khashiba in Sinai. The key result was the phased withdrawal, including the uprooting of new settlements, of Israel from Sinai over a three-year period. The new border was to be an open border between Israel and Egypt.

Although the Israeli–Egyptian border was restored to its original position, Israel retained Wadi Taba, immediately south of the Israeli part of Eilat, and built a luxury hotel on the disputed territory. The area involved was only 900 m^2, but Israel and Egypt could not agree and the case went to arbitration in 1986. The tribunal found in favour of Egypt.

Sinai is vital in guarding the approaches to the Suez and Egypt. It is an historic battleground and has witnessed conflict from 1500 BC to 1973. Economically, it has some significance for oil production and tourism.

Status

Since 1979, there have been very few incidents in the area and the situation is considered stable. More importantly, Israel and Egypt have managed to retain diplomatic relations since 1979. Initially, Egypt was ostracized by the Arab world for the Camp David Accords, being ejected from the Arab League whose headquarters were removed from Cairo, but under President Mubarak it had, by the end of the 1980s, regained its leading role. This was, if anything, reinforced during the Gulf Conflict.

The 1979 treaty divided Sinai into three zones. Zone C, parallel to the border and approximately 50 km wide, was allowed no Egyptian troops. The central zone, B, with an average width of 100 km, was allowed no more than 4,000 lightly armed Egyptian troops, whereas Zone A, in the west, was allowed up to 22,000 troops. On the Israeli side of the border, Zone D, a strip 5 km deep, was permitted a limit of 4,000 Israeli troops. The United Nations force in Sinai amounts to some 2,600 troops.

References

Blake, G.H., Dewdney, J. and Mitchell, J. (1987), *The Cambridge Atlas of the Middle East and North Africa*, Cambridge University Press, Cambridge.

Blake, G.H., and Drysdale, A. (1985), *The Middle East and North Africa: A Political Geography*, Oxford University Press, Oxford.

Day, A.J. (ed.) (1984), *Border and Territorial Disputes*, Longman, London.

Downing, D. (1980), *An Atlas of Territorial and Border Disputes*, New English Library, London.

Dowman, I. (1990), 'The Taba Case: A Cartographer's Perspective' in N. Beschorner, St. J.B. Gould, and K. McLachlan (eds), *Sovereignty, Territoriality and International Boundaries in South Asia, South West Asia and the Mediterranean Basin*, Proceedings of a seminar held at the School of Oriental and African Studies, University of London, pp. 17–21.

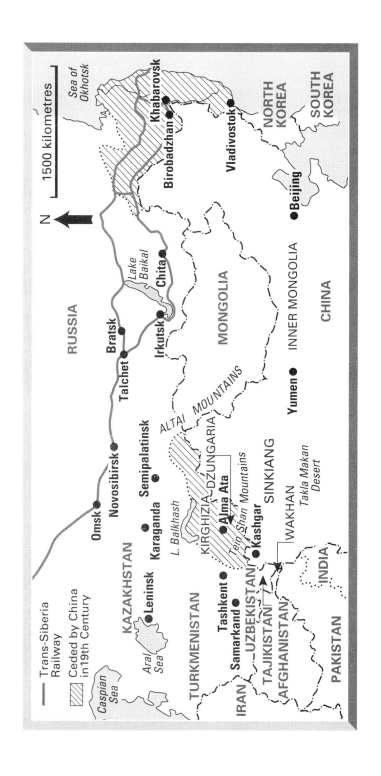

62 The Sino-Russian (formerly Soviet) Border

Description

Forming an arc from the Pamir plateau of Central Asia to the Pacific Ocean, the 6,640-km boundary is divided into two sections by Mongolia. The far eastern (Manchuria/Siberia) sector extends for some 3,700 km, primarily along the courses of the rivers Argun, Amur and Ussuri. The Central Asian (Sinkiang/Kazakhstan, Kirghizia, Tajikistan) sector is some 3,000 km in length and crosses some of the highest mountains in the world. From the 16th to the 19th centuries, Russia and China had gradually subdued the whole of northern and northeastern Asia before, in the 1860s, the two empires clashed.

History and importance

With reference to the far-eastern sector, the Manchu conquest of China was completed between 1644 and 1662, but the thinly populated areas north of the Amur River and east of the Ussuri River were never effectively controlled. Meanwhile, in 1644, a Russian military expedition reached the mouth of the Amur River and, in 1665, an outpost was established there. This led to sporadic fighting, but in 1689, the Treaty of Nerchinsk fixed the boundary between the Chinese and Russian Empires north of the Amur, along the line of the Stanovoi mountains.

By the middle of the 19th century, China was weakened by two Opium Wars and the Taiping Rebellion, and Russia advanced its settlement. As a result, in 1858, the Chinese commander in the Amur area was forced to sign the Treaty of Aigun, which gave Russia sovereignty over some 480,000 km^2 north of the Amur River, while a further 330,000 km^2, east of the Ussuri, were placed under joint Sino-Russian sovereignty. Later in the year, at the Treaty of Tientsin (Tianjin) these changes were confirmed. At the following Treaty of Peking in 1860, the Tsar gained the area east of the Ussuri and defined the Central Asian boundary.

During the 17th century, Chinese control was established in Sinkiang (Xinjiang), but there was little settlement and there were several armed uprisings. The Treaty of Peking fixed the boundary, but at the subsequent Treaty of Chuguchak (1864) a Russian claim to 900,000 km of territory that had not been under China's effective control was accepted. In the same year, there was a revolt throughout the whole of Sinkiang and to prevent this spreading to his own Central Asian lands, the Tsar aided China.

By 1871, Russia had occupied the Ili and Tekkes valleys up to Kulja, and by 1877, with Russian aid, Sinkiang was finally reconquered. At the Treaty of St. Petersburg (1881), Russia agreed to evacuate half the territory it had occupied in return for the retention of the other half and territorial concessions further north, to the east of Lake Zaysan. Thus, by the end of the 19th century, with Russian expansion into the Pamir Knot, the delimitation of the Sino-Russian border was complete.

With the overthrow of the Chinese Empire in 1911, the new republic demanded the abrogation of all the 'unequal treaties', but before any action could be taken, the Bolshevik Revolution occurred in 1917. On 25 July 1919, with the Karakhan Declaration, the Soviet government stated that past treaties were null and void and it renounced seizures of territory. However, it did not mention the treaties of Aigun, Peking or Chuguchak. On 31 May 1924, there was a Sino-Soviet Agreement to annul the treaties and redemarcate but pending its completion, to maintain present boundaries.

China was then embroiled in 20 years of civil war and an invasion by Japan, and Sinkiang passed to the Soviet sphere of influence. Conflict ceased with the Communist victory in 1949, and in 1950 a 30-year Treaty of Friendship, Alliance and Mutual Assistance was signed with the Soviet Union. No mention was made of the borders. However, various official Chinese maps of the 1950s showed boundaries at variance with those previously agreed.

A map of 1953 indicated a Sinkiang–Tajikistan boundary in the Pamirs, several hundred miles to the west of its official position.

From the 1960s until recently, relations between the two premier Communist states deteriorated. Border incidents began in July 1960 and, during that year, the Soviet Union alleged there were 5,000 Chinese border violations. On 8 March 1963, China raised the question of nine 'unequal treaties' and in September 1964, tension mounted along the Sinkiang border and about 50,000 Kazakhs fled to the Soviet Union. Negotiations began on 25 February in Peking, but by May were abandoned as the Chinese delegation laid claim to over 1.5 million km^2 of Soviet territory.

In 1966, accusations from both sides continued and tension mounted when the Cultural Revolution began. On 2 October, it was alleged that over two million Chinese were involved in mass demonstrations on the border. Between 1967 and 1969, there were frequent clashes, particularly on an uninhabited island (Damansky or Chenpaer) in the Ussuri River. These clashes occurred 180 km south of Khabarovsk and some 400 km north of Vladivostok. The status of islands in the Ussuri was in fact disputed in the treaties of Aigun and Peking. The Soviet Union produced a map of 1861, showing the boundary line on the Chinese bank of the Ussuri, but China maintained that this was imposed and the correct boundary should be a median line.

By far the worst year for clashes was 1969. On 2 March, a Chinese ambush of Damansky Island left 31 Soviet soldiers dead and 14 wounded. Fighting escalated on 15 March, when the Soviets captured Damansky Island. In April, fighting moved west to Central Asia near Chuguchak where violent clashes continued into July. Between 18 June and 8 August, a Joint Sino–Soviet Commission on Navigation on the Far Eastern frontier rivers met in Khabarovsk and agreement was reached. Claims and counter-claims followed over hundreds of islands and an area of at least 1,000 km^2 and by 23 August, China issued a statement that war might break out at any time. This was defused on 11 September when Soviet Prime Minister Alexei Kosygin visited Peking.

It was estimated at the time that there were some 270,000 Soviet troops available in the immediate vicinity of the border. These were opposed by at least a million Chinese troops. Later in the year, border negotiations opened in Peking to continue virtually without progress until the end of 1978.

In the meantime, Chinese officials in Central Asia declared that 20 areas in Sinkiang, varying in size from 1,000 to 30,000 km^2, were in dispute. In 1977, the dispute over one island at the junction of the Amur and Ussuri Rivers was resolved, but further talks in 1979 were inconclusive.

On 16 June 1981, a new source of friction emerged when the Soviet–Afghanistan border treaty recognized as Soviet an area northeast of Afghanistan to which China laid claim. This treaty proved to be a prelude to the Soviet occupation of the Wakhan Salient (panhandle). On 22 July, China stated that the Soviet claim under the treaty was 'illegal and invalid'. The dispute in fact concerned some 20,000 km^2 Russia had occupied during the 1890s.

Since the border was based on 19th-century Manchu–Tsarist treaties, China maintained they were 'unequal'. With some 1.5 km^2 at stake, including the Russian outlet to the Pacific, the port of Vladivostok and a part of the trans-Siberian railway, the issue must be taken seriously.

Status

With *glasnost* and the subsequent collapse of the Soviet state, there have been significant changes in Sino-Russian relations. There are now regular meetings about border concerns and a new agreement was reached on 16 May 1991. Since the border issues now concern not only the Russian Federation, but also the emergent states of Kazakhstan, Kirghisia and Tajikistan of the CIS, the position is immensely complicated. It is clearly difficult to predict what will happen, but settlement without conflict seems unlikely.

At the moment, China has 1.4 million troops and over 5,000 aircraft stationed near the border. These are faced by 460,000 troops of the former Soviet Red Army and 2,500 aircraft nominally at least under Russian control. A change of regime in China seems at present unlikely, at least in the medium term, and therefore resentment over the 'unequal treaties' appears set to continue. However, the economic condition and uncertain foreign policies of the newly independent former Soviet states could engender sudden and unpredictable changes, like demilitarization, that could greatly influence the situation.

References

Boundary Bulletin, No. 2, (1991), International Boundaries Research Unit, Durham University.

Boyd, A. (1991), *An Atlas of World Affairs*, Routledge, London.

Day, A.J. (ed.) (1984), *Border and Territorial Disputes*, Longman, London.

Downing, D. (1980), *An Atlas of Territorial and Border Disputes*, New English Library, London.

The Economist, (1989), 4 February.

Karan, P.P. (1964), 'The Sino-Soviet Border Dispute', *Journal of Geography*, **LXIII**, No. 5, pp. 216-22.

Levine, S.I. (1987), 'Sino-Soviet Relations in the Late 1980's: An End to Estrangement?' in L.E. Grinter and Y.W. Kihl (eds), *East Asian Conflict Zones*, Macmillan, London, pp. 29-46.

United States Department of State (1974), *China - U.S.S.R, Boundary*, International Boundary Study No. 64 (revised), Office of the Geographer, Bureau of Intelligence and Research, Washington DC, 22 January.

63 South Lebanon

Description

Lying on the eastern shore of the Mediterranean, at the historical crossroads of the world, Lebanon has been the scene of almost continuous conflict since shortly after independence in 1943. The country has an area of 10,452 km² and an estimated population of 3.4 million. Despite its small size, it includes a marked variation in relief and, more importantly, an extremely complex social mosaic. Some 93 per cent of the population is Arab, but there are Armenian and Kurdish minorities and four languages are in common use. However, the major difficulties are related to religion. The population is approximately 60 per cent Muslim and 40 per cent Christian, but the Muslims are divided into Shia, Sunni and Druse in the approximate ratio of 6:4:1 respectively and the Christians include Maronites (25 per cent), Greek Orthodox (7 per cent), Armenian Orthodox (5 per cent) and Greek Catholic (4.5 per cent). Secure in the mountains, the Maronites were able to resist the 7th-century Islamicization of the country and, with the trading links they developed, by the 19th century were able to dominate commerce. The other distinctive community is that of the Druse which developed from Islam in the 11th century.

A fundamental contribution to the problems of Lebanon is the variety of religions and sects, but the situation was exacerabted by French colonial rule. In 1860, following a Maronite massacre by the Druse, France established a protectorate and the Maronite area in the region of Mount Lebanon was granted autonomous status. After World War I, France was given the League of Nations' mandate for Syria, which was divided into three Arab states and the Christian state of Great Lebanon. In 1932, a census, conducted by the French, indicated that within the semi-autonomous republic of Lebanon there was a majority of Christians (55 per cent). This result was used in 1943 to develop the 'National Pact' as the basis for power sharing between Christians and Muslims in a ratio of 6:5. Moreover, it was agreed that the offices of state would be awarded according to religion. The president would be a Maronite, the prime minister a Sunni Muslim and the president of the National Assembly would be a Shia Muslim. Even if the original census results can be accepted as valid, such a rigid structure was bound to result in complications. Furthermore, since the arrangement was agreed in 1943, the population has undergone almost constant change, which has produced an ever-increasing number of Muslims.

The Arab–Israeli confrontation has been the other major influence producing tensions in Lebanon. After the wars of 1948–9 and 1967, large numbers of Palestinian refugees settled in Lebanon, altering the delicate population balance. Their arrival brought southern Lebanon into prominence, a prominence reinforced by the establishment of the Palestine Liberation Organization (PLO) headquarters in Lebanon, following its expulsion from Jordan in 1970.

History and importance

Although Lebanon itself has witnessed communal strife over a longer period, the history of South Lebanon as a distinctive entity and battleground dates effectively from about 1970. From then until the major Israeli invasion of 1982, the area was enveloped in almost continuous turmoil, fuelled by the Arab–Israeli conflict, but underpinned by social differences within Lebanon itself. By 1975, there were some 400,000 Palestinians, mostly in camps in South Lebanon, attracting some support from fellow Arabs, but hostility from the Maronites. Palestinian cross-border raids brought Israeli reprisals and the country became increasingly less stable. In 1973, the Maronite militia came into conflict with the Palestinian guerrillas, and in the following few years a number of other extreme organizations joined the fray. In 1974, the Shia Muslims began to organize Amal (a Shiite organization formed in 1968) groups and by mid-1975, there was full-scale civil war.

At first, it appeared that the Druse–Palestinian alliance would prevail, but in the end, the decisive influence was that of Syria, which took control and imposed a provisional settlement. Having supported the non-Muslim side, Syria was reviled by the Arab world and was forced to enter negotiations. These resulted in the withdrawal of the armed PLO units from Beirut, but in the south, fighting restarted between Palestinians and Christians. Responding to this and PLO attacks across the border, Israeli forces invaded as far north as the Litani River. At the subsequent cease-fire (July 1978), the key positions were handed over to the Lebanese Christian militia and a semi-independent Christian state of Free Lebanon was formed. Neither the Lebanese army nor the United Nations Interim Force in Lebanon (UNIFIL) proved able to resist the establishment of what was effectively an Israeli puppet state.

Israel was thus actively promoting the disintegration of Lebanon and, in particular, the establishment of its own interests in the southern third of the country. The most obvious reason for this was to provide a buffer zone, separating Israel from the warring factions in Lebanon proper and ensuring the security of northern Galilee. Furthermore, the Zionist movement had long claimed South Lebanon as part of Israel. The other factor suggested as a possible Israeli motive concerned the water balance. By advancing to the Litani River, Israel provided itself with a potential new water supply and there has been continuing speculation about the possible diversion of the Litani into the Jordan system.

Although the troops had been withdrawn in 1979, attacks continued on Palestinian positions throughout 1980 and 1981, until a cease-fire was agreed (24 July 1981). This lasted for ten months, after which, following a particularly savage exchange, Israeli forces once more invaded (6 June), this time invading as far as west Beirut. The main aim of the operation, named 'Peaceful Galilee', was the removal of the PLO and this was largely accomplished, following the agreement of 19 August 1982 when the Organization was forced to withdraw from Beirut. However, the situation immediately deteriorated, following the massacre of Palestinians in the Chattila and Sabra refugee camps by phalangist (Christian) militia, with the apparent collusion of the Israeli forces.

In the face of growing international criticism, Israel offered to withdraw in the context of a peace agreement, which would guarantee a demilitarized zone adjacent to the northern border of Israel. Negotiations opened at Khalde and an agreement was reached on 17 May 1983 for:

(a) the withdrawal of Israeli forces;
(b) the termination of the state of war between Israel and Lebanon; and
(c) the establishment of a security zone in South Lebanon.

The agreement was never ratified, but Israel mounted a staged withdrawal from Lebanon, which was completed on 6 June 1985. However, a security zone extending some 10 km from the border and patrolled by the Israeli-backed South Lebanon Army (SLA) was established. Furthermore, as indicated by its actions over the next few years, Israel maintained a right of re-entry into the zone.

The importance of South Lebanon is that it constitutes the one almost continuously active front between Israel and the Arabs. There is also the question of the water balance within the Jordan Basin, but any interference with the flow of the Litani would be geopolitically extremely risky. In the long term and perhaps even more important than either of these considerations is the fact that South Lebanon presents Israel and, to an extent, Syria with a bargaining opportunity. Complete Israeli withdrawal of influence might be linked with the withdrawal of Syrian forces in the north. There must also remain a possibility that South Lebanon might be traded for the occupied territories in the Golan Heights.

Status

With the ending of the civil war in Lebanon, attention has concentrated on South Lebanon and the possible implementation of UN Security Council Resolution 425 under which the area was to be returned to Lebanese government control. The number of serious incidents has also declined and only two of the main militias, the Hezbollah, supported by Iran, and the SLA, sponsored by Israel, have refused to disarm. Furthermore, the military capability of the PLO has been greatly weakened. However, while all these changes may appear to make the implementation of Resolution 425 easier, they also reduce any pressure on Israel to agree to such a permanent settlement.

At present, Israel effectively controls the zone,

there are few incidents and there remains the potential to obtain additional water supplies. With the reduction in tension, these advantages are all obtained at a modest cost. The main hope for a solution to the problem must depend upon a successful outcome to the current peace talks. Meanwhile, South Lebanon remains a key world flashpoint.

References

Boyd, A. (1991), *An Atlas of World Affairs*, ninth edition, Routledge, London.

Calvocoressi, P. (1991), *World Politics Since 1945*, sixth edition, Longman, London.

Rapoport, D.C. (ed.) (1988), *Inside Terrorist Organizations*, Frank Cass, London.

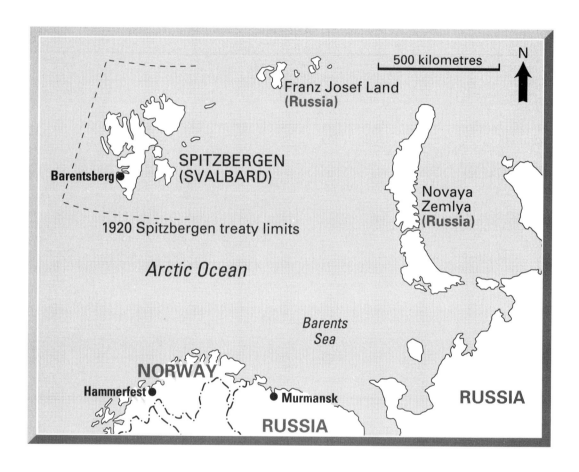

64 Spitzbergen (Svalbard)

Description

An archipelago, consisting of several large and many small islands, Spitzbergen lies 400 nml north of the Norwegian mainland. On 9 February 1920, Norway gained undisputed sovereignty of the islands by means of a treaty signed by 13 countries and, later, acceded to by 36 states, including the Soviet Union. The Treaty confers upon Norway full sovereignty over all islands within an area bounded by the latitudes 74° and 81°N and the longitudes 10° and 35°E. However, all parties to the Treaty enjoy equal fishing and hunting rights on the islands and in the territorial waters and have equality in all maritime, industrial, mining and commercial activities. All activities are, nevertheless, limited to peaceful uses. Thus in the late 1940s, Norway was able to resist Soviet pressure to station a garrison in Spitzbergen.

The population of the islands comprises approximately 2,000 Russians, based at Barentsberg and Pyramiden and 1,000 Norwegians, centred at Longyerbyen, both communities being concerned with coal mining. Each operation yields approximately 450,000 tonnes of coal per month.

History and importance

The Treaty of 1920 made no provision relating to the continental shelf outside territorial waters, and the total area, including Spitzbergen itself, covers slightly less than 400,000 nml^2. In 1970, Norway asserted jurisdiction over the shelf surrounding Spitzbergen, claiming it as a natural prolongation of Norway's mainland continental shelf. In contrast, the Soviet Union maintained that Spitzbergen had its own continental shelf to which the Treaty should apply.

In 1977, Norway extended its jurisdiction by introducing a fisheries protection zone as part of a Norwegian EEZ. The effect of this 200 nml EEZ extension is that all sailing routes to the Russian northern ports pass through waters under Norway's jurisdiction. On several occasions, as a result, the former Soviet Union attempted to persuade Norway to accept joint administration. Had this been granted, the overwhelming power of the Soviet Union could have given it virtual control of the islands and associated strategically vital northern waters. Indeed, during the 1970s there was mounting Soviet military activity in the area and numerous violations of the 1920 Treaty occurred. In fact, in defiance of Norwegian sovereignty, the Soviet communities on the islands became virtually autonomous enclaves.

Given the overwhelming strategic importance of the location, the level of conflict has been relatively minor. The mining of coal is seen as a factor of little consequence other than a pretext for allowing Russia to maintain a presence for potential military purposes. However, the potential for fishing and the exploitation of the continental shelf for petroleum is of rising significance.

Status

Although the key territorial issues remain unresolved, the demise of the Soviet Union and the consequent decline in East–West tension has resulted in a reduction in the strategic significance of the region. The way in which Russia will develop the Kola Peninsula and what will be the new requirement for the Northern Fleet must remain a matter of conjecture. If the former Soviet military retain a high profile in Russian society, the potential of the Northern Fleet may be little diminished. On the other hand, defence cuts being currently proposed would seem to indicate that, given the state of the Russian economy, it will be impossible to retain such a mighty military presence.

References

Armstrong, T., Rogers, G. and Rowley, G. (1978), *The Circumpolar North*, Methuen, London.
Boyd, A. (1991), *An Atlas of World Affairs*, Routledge, London.
Leighton, M.K. (1979), *The Soviet Threat to N.A.T.O.'s Northern Flank*, Agenda Paper No. 10, National Strategy Information Center, Inc., New York.
Luton, G. (1986), 'Strategic Issues in the Arctic Region' in E.M. Borgese and N. Ginsburg (eds), *Ocean Yearbook* 6, University of Chicago Press, Chicago, pp. 399–416
Prescott, J.R.V. (1985), *The Maritime Political Boundaries of the World*, Methuen, London.

65 The Spratly Islands

Description

The Spratly group is located approximately 300 nml west of the Philippine island of Palawan, 300 nml east of Vietnam and 650 nml south of Hainan (China). It comprises a vast expanse of reefs, shoals and islands, known to the Chinese as Mansha and to the Vietnamese as Truong Sa. Western navigators have divided the area into the Spratly Islands, the Dangerous Ground and the Reed Bank, but the whole area is usually referred to as the Spratly Islands. The area claimed by the Philippines is called Kalayaan or Freedom Land. The Four Claim Area covers approximately 70,000 nml^2 and ranges from 7–12°N and 11–118°E. The islands are very small; the largest, Itu Aba, is only 960 by 400 m (36 hectares). It rises just over 2 m above the water.

All claimants cite geographical proximity as the factor in their favour. China, Taiwan and Vietnam claim entirely according to distance. The claim of the Philippines follows the 1956 Proclamation, by which an attempt was made to establish the independent state of Freedom Land in the Spratlys. This claim was laid on 11 June 1978. The claim of Malaysia is based on geology in that the islands stand on the Malaysian Continental Shelf. There are, as a result, overlapping claims, with all the islands claimed by at least two states. No island is claimed by more than four states.

History and importance

In 1933, France annexed the Spratly Islands as part of its larger claim to Indonesia. China and Japan protested, the latter claiming continuous commercial occupation since 1917. In 1939, Japan annexed the islands, but in 1946 a Chinese (Kuomintang) naval expedition took possession and garrisoned Itu Aba. At the Treaty of San Francisco in 1951, Japan renounced its claim, but in 1954, after the division of Vietnam, South Vietnam lodged a claim.

In May 1956, the government of the Philippines made its first claim and later that year, in August, a South Vietnamese garrison was established in the islands. In July 1971, a note was sent from the government of the Philippines to South Vietnam, demanding withdrawal, after an attack upon a Philippine fishing boat. At that stage, several islands were occupied and in September 1973, a decree was promulgated by South Vietnam, incorporating the Spratlys into its adjacent province. However, on 11 January 1974, China protested and expelled the South Vietnamese from the Paracels. The South Vietnamese response was to occupy three of the Spratlys.

In April 1975, with the fall of Saigon, Communist Vietnamese troops took over the Spratly garrisons. When, in May 1976, the government of the Philippines announced that a joint Swedish–Philippines consortium was to start exploration for oil in the area, the provisional revolutionary government (South Vietnam), China and Taiwan, protested. With the reunification of Vietnam, in July 1976, there was a confrontation between Hanoi and Beijing over the Spratlys. In March of the same year, the Philippines established a military command on Palawan and by March 1978 had set up garrisons on seven islands.

In January 1978, there was a Philippines–Vietnam agreement to settle the dispute 'in a spirit of conciliation and friendship'. In March, a similar agreement was reached with China, but on 11 June, the Philippines claim was made. At the same time, Malaysia's first claim was laid. This was to the island of Amboyna Cay and was justified by the fact that a survey had found it to be in Malaysian waters. However, it is not waters that give title to islands, but islands to waters. Malaysia bolstered its claim by constructing obelisks on the Louisa and Commodore Reefs. Nevertheless, the monument on the Commodore Reef is believed to have been destroyed by the Philippines. On 14 September 1979, Philippino President Fredinand Marcos clarified the position by stating that the Philippines'

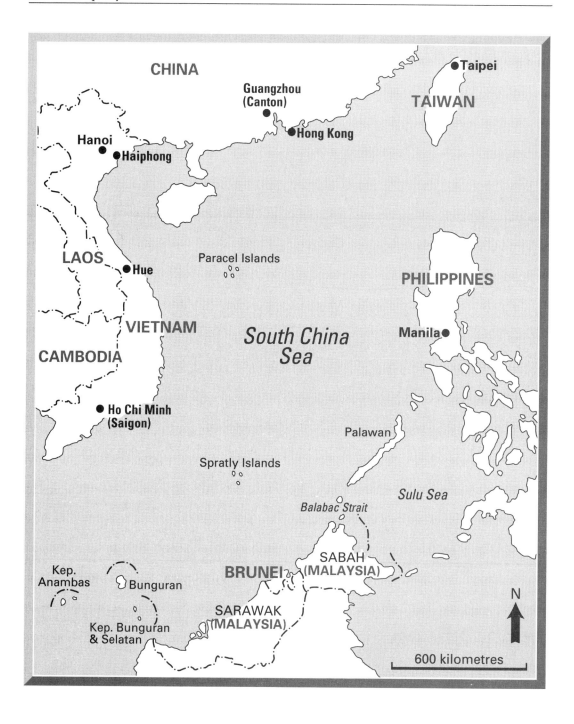

claim was only to the seven islands occupied. However, in August 1980, the foreign minister of Vietnam protested at the Philippines' occupation. On 10 April 1983, a German yacht was sunk by fire from one of the garrisoned islands.

None the less, in 1988, these claims and counter-claims were partly resolved by action, when forces from China landed on Fiery Cross and Curteron Reefs. An island-hopping race with Vietnam began. On 14 March, the Chinese and Vietnamese forces clashed near Chigua Reef and three Vietnamese ships were damaged and two sunk. China was accused of obstructing rescue attempts and the Vietnamese were alleged to be 'continuing to seize Chinese islands on the pretext of carrying out rescue operations'. By April, China had control of 6 islands and reefs, and Vietnam, an extra 15 making a total of 21. A Vietnamese call for negotiations was rebuffed and in November it was alleged that a Chinese destroyer had fired on a Vietnamese ship near the Collins Reef.

During these clashes, in April 1988, Taiwan stated that it would 'fight against Vietnam' with China, if necessary, to maintain the Chinese claim. On 13 April 1990, a Chinese proposal was made for the joint development of the Spratly Islands: 'China is ready to join efforts with Southeast Asian countries to develop the Spratly Islands while putting aside for the time being the question of sovereignty'. This would, of course, entail demilitarization and joint exploration and it would require Vietnam to give up its superior military position. Thus, co-operation seems unlikely.

In May 1991, the position became even more complex when Malaysia announced the development of Terumbu Layang Layang Atoll (Swallow Reef) as a tourist resort. This brought forth storms of protest and, as a result, Vietnam reaffirmed its sovereignty in April and in May, while China increased its military activity.

The Spratlys are particularly important strategically in that they command the main shipping lanes from Singapore to Japan. Furthermore, it is reported that the presence of oil in the Reed Bank area has been confirmed by the Philippines. The islands themselves are of minimal economic value, but they allow sea space claims of the order of 150,000 nml^2 .

Status

The seemingly inconsequential Spratly Islands, virtually unnoticed until the 1930s, now furnish what many consider to be the key potential flashpoint in Southeast Asia and possibly, the entire Pacific area. Sporadic conflict continues and, with five states claiming partial jurisdiction and two undertaking active development of their claims, no solution is likely. This is particularly so given the diverse ideological, economic, ethnic and religious characteristics and aspirations of the states involved. Presently, all five states occupy some territory in the group.

References

Boundary Bulletin, No. 1, (1991), International Boundaries Research Unit, Durham University.
Boundary Bulletin, No. 2, (1991), International Boundaries Research Unit, Durham University.
Boundary Bulletin, No. 3, (1992), International Boundaries Research Unit, Durham University, January.
Boyd, A. (1991), *An Atlas of World Affairs*, Routledge, London.
Day, A.J. (ed.) (1984), *Border and Territorial Disputes*, Longman, London.
Dzurek, D.J. (1985), 'Boundary and Resource Disputes in the South China Sea' in E.M. Borgese and N. Ginsburg (eds), *Ocean Yearbook 5*, University of Chicago Press, Chicago, pp. 254-84.
The Economist (1989), 21 May.
Leng, L.Y. (1989), 'The Malaysian – Philippine Maritime Dispute', *Contemporary Southeast Asia*, 11, No. 1, pp. 61-74.
Park, C. (1980), 'Offshore Oil Development in the China Seas: Some Legal and Territorial Issues' in E.M. Borgese and N. Ginsburg (eds), *Ocean Yearbook 2*, University of Chicago Press, Chicago, pp. 302-16.
Prescott, J.R.V. (1985), *The Maritime Political Boundaries of the World*, Methuen, London.
Thomas, B.L. (1989), 'The Spratley Islands Imbroglio: A Tangled Web of Conflict' in *International Boundaries and Boundary Conflict Resolution*, 1989 Conference Proceedings, International Boundaries Research Unit, University of Durham, pp. 413-25.
Weatherbee, D.E. (1987), 'The South China Sea: From Zone of Conflict to Zone of Peace?' in L.E. Grinter and Y.W. Kihl (eds), *East Asian Conflict Zones*, Macmillan, London, pp. 123-48.

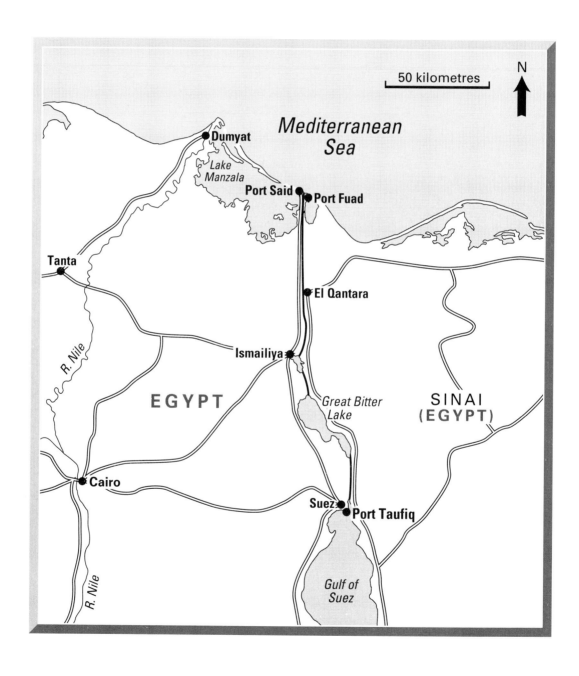

66 The Suez Canal

Description

Suez is a sea-level canal, linking the Mediterranean to the Red Sea, by way of the Gulf of Suez. The concept of a waterway through the Isthmus of Suez goes back almost four thousand years, but it was not until 1854 that the French engineer Ferdinand de Lesseps was contracted to construct the canal. In 1869, the 165.8 km long Suez Canal, was opened. The route through the sandy isthmus was chosen to take advantage of several marshy areas and shallow lakes. Originally 21.9 m wide and 7 m deep, by 1956, the maximum permissible draft had risen to 10.7 m and this was increased in 1961 to 11.3 m, in October 1980, to 16.1 m and then to the present maximum of 19.5 m. From October 1980, the width was increased to 270 m, with the result that the canal could then take tankers up to 150,000 tonnes, fully laden. Phase two, postponed as a result of the world shipping recession and the decline in oil consumption, would have produced an increase in width to 315 m and in depth to 23.5 m, thereby allowing the passage of ships of 250,000 tonnes. The average transit time is 24 hours and some 80 ships per day use the canal.

History and importance

The earliest canal, an improved wadi or natural channel, running from Lake Timsah to the Nile delta, appears to have been in use from about 2000 BC. However, it was not until the official opening of the Suez Canal on 17 November 1869 that a complete sea-level waterway was available for ocean-going ships. In 1875, Britain attained a half share of the Suez Canal Company and in 1882, largely to ensure its control of the canal, occupied Egypt. Britain's position was recognized internationally by the Constantinople Convention of 1888. Thus, Britain held a series of strategic locations which linked Europe with India: Gibraltar, Malta, Suez and Aden. Suez became known as the 'imperial lifeline' and its protection became a major British preoccupation.

Therefore, in 1956, following the nationalization of the Canal Company by President Nasser, Britain and France were prepared to go to war to protect their interests. However, the hostilities were opposed by the two superpowers and the campaign of October 1956 quickly ended, but the canal itself was blocked by wrecked ships. It was re-opened in April 1957, but was again closed in June 1967, as a result of the Six-Day War, when Israel invaded Sinai. This time, political problems and clearance of the canal took longer and it was not until June 1975 that the canal was re-opened to international, including Israeli shipping.

The Suez Canal route drastically reduced the distances and therefore fuel consumption of ships transiting from London to the Persian/Arabian Gulf and the Far East. The saving in distance to Gulf Kuwait and Bombay is approximately 42 per cent and that to Singapore, 30 per cent. Thus, the closure from June 1967 to June 1975 required a radical reorganization of the infrastructure of world shipping. The major results were the construction of the SUMED pipeline from the Gulf of Suez to Alexandria, thereby avoiding the Suez Canal, and the development of the Cape Route round the Cape of Good Hope. The other key factor during the period was the emergence of super-tankers, designed to combat the extra costs of the Cape Route, but too large to transit Suez.

The overall result of its closure after the Six-Day War was therefore that the Suez Canal lost considerably in status. For example, in June 1975, only 20 per cent of the world's tankers could use the canal. However, Suez is now recovering and ships are being built to its maximum specification. Furthermore, new techniques for the passage, such as a partial unloading and reloading have been developed. None the less, the importance of the Cape Route as a safeguard remains.

By far the most important cargo transiting the

canal is oil and oil products which accounts for over 29 per cent of the total of 271.8 million tonnes. Owing to the new techniques employed, the passage of Very-Large and Ultra-Large Cargo Carriers (VLCCs and ULCCs) has risen from 400 vessels of over 200,000 dwt in 1989, to in 1990, 518 of which 142 were over 300,000 dwt. However, the volume of trade has still not reached that achieved before the closure, and the figure for 1990 for north-bound goods was still 23.7 million tonnes short of that for 1966. On the other hand, the number of container ships has increased fourfold since 1977 to a 1990 figure of 3,077 ships. Bulk carriers and conventional cargo vessels have both declined somewhat in the late 1980s and early 1990s, but the number of warships more than doubled from 1989 to 1990.

The major north-bound cargoes were (1990): oil and oil products (65.8 million tonnes), coke and coal (12.1 million tonnes) and ores and metals (10.3 million tonnes). The key items of south-bound traffic were: fertilizers (14.7 million tonnes), oil and oil products (13.8 million tonnes), fabricated metals (13.1 million tonnes) and cereals (9.9 million tonnes). The major users situated north of the canal were Italy (13.2 per cent) and the United States, the former Soviet Union and the Netherlands (all just under 7 per cent). The greatest users from south of the canal were Saudi Arabia (15.2 per cent), India (9.3 per cent), Egypt (6.9 per cent), Australia (6.5 per cent) and China (6.2 per cent).

The construction of the SUMED pipeline, a 336-km-long double pipeline, bypassing Suez and opened in 1975, has, of course, reduced trade. Indeed, the SUMED pipeline has now surplanted the Suez Canal as the main oil routeway from the Red Sea to the Mediterranean. In 1990, 80 million tonnes of crude oil were pumped through the pipeline and the figure is expected to rise to 120 million tonnes by the end of 1992. According to SUMED pipeline studies, another 80 million tonnes are transported by the Cape Route and it is considered that, of this, the SUMED pipeline could win some 35–40 million tonnes. The main user is Saudi Arabia, which, in the first half of 1991, accounted for 80 per cent of the traffic. Even greater strains may be placed on the canal as there is potential for a further pipeline, which would bring the capacity of the SUMED pipeline to 240 million tonnes per annum.

Status

The Suez Canal is gradually recapturing some of its former status, but it is unlikely ever to regain its full supremacy over the Cape Route. However, some 41 per cent of Western Europe's oil comes from the Gulf region and this figure is likely to increase in the future. Since the states of the Gulf region account for some 65.6 per cent of the world's proved oil reserves, they will increasingly dominate world supplies.

However, the canal, having been closed twice, remains vulnerable. From 1 September 1970, during Operation Desert Shield, a Hezbollah group, with a shoulder-launched missile, was intercepted before it could attack a transiting ship. Given its vital importance in the world's infrastructure, particularly for oil, which is the leading commodity traded, the Suez Canal is always likely to remain a potential flashpoint.

References

Blake, G.H., Dewdney, J. and Mitchell, J. (1987), *The Cambridge Atlas of the Middle East and North Africa*, Cambridge University Press, Cambridge.
Lapidoth, R. (1975), *Freedom of Navigation With Special Reference to International Waterways in the Middle East*, The Hebrew University, Jerusalem.
Mideast Mirror (1991), 5, No. 226, 20 November.
Times Atlas of the Oceans (1983), Times Books, London.

67 Surinam

Description

Surinam, the former Dutch Guiana, in many respects closely resembles its neighbour, Guyana. The country is largely densely forested and the majority of the population live on a narrow coastal strip where the chief hazards are potential sea-level rise and saline incursions. The population is 380,000 and this comprises the following ethnic mix: Asian-Indian, 35 per cent; Creole, 32 per cent; Indonesian, 15 per cent; Bush Negro, 10 per cent; Amerindian, 3 per cent and Chinese, 3 per cent. Religious adherence presents an equally complex mosaic: Hindu, 27 per cent; Roman Catholic, 23 per cent; Muslim, 20 per cent and Protestant, 19 per cent.

The economy is completely dominated by the bauxite industry which accounts for 80 per cent of export earnings and 40 per cent of tax revenues. Furthermore, bauxite (an ore of aluminium) is considered at least marginally strategic and Surinam is a major supplier to the United States. Politically, the military remain dominant and Dutch aid was frozen after a military coup in December 1990.

History and importance

Like Guyana, Surinam has boundary disputes in the east and in the west. That in the west has already been discussed (Map 27). That in the east with French Guyana, concerns an area between the Litani and Marouini Rivers, both of which are tributaries of the Marowijne River (known as the Maroni in French Guyana, and in both countries as the Lawa). The area at issue is some 5,000 km² of thickly forested, mountainous terrain which rises to form the Tumac-Humac (Tumucumaque) range, the common border with Brazil.

In 1680, the French and Dutch agreed on the Maroni (Marowijne in Dutch) River as the boundary between their territorial claims in the region, but this was only navigable for some 40 km inland and neither country was interested in the interior. In the early-19th century, differences arose as to whether the boundary followed the Tapanahoni (now in mid-Surinam) or Awa River.

The Portuguese–French Treaty of 1815 and Convention of 28 August 1817 stated that the western boundary of French Guiana was the River Tapanahoni, but the Netherlands protested. In 1861, a mixed commission was unable to decide which river constituted the Upper Maroni and discussions lapsed until 1876.

On 29 November 1888, in Paris, the Netherlands and France signed an undertaking to submit to arbitration, the selected arbitrator being Alexander III, Tsar of Russia. Between 13 and 25 May 1891, the arbitration body sat and eventually determined in the Netherlands' favour. The Awa was defined as the upper course of the Maroni. Later, rapid development led to further disputes over the tributaries, and a conference on the dispute (25 April–13 May 1905) recommended that the boundary should be the thalweg in the Maroni estuary and along the Maroni, Awa and Itany (Litani) Rivers.

On 13 September 1915 this was modified when an agreement between the Netherlands and France substituted a median line for a thalweg in the estuary of the Maroni. Despite the recommendation of 1905, however, the area between the Litani and the Marouini was still claimed by the Netherlands and, since independence in 1975, by Surinam.

Negotiations in November 1975 and February 1977 formulated a treaty whereby Surinam would recognize French sovereignty over most of the area in return for 500 million francs' development aid for joint development of the mineral wealth of the region. This treaty was initialled on 15 August 1977, but despite further meetings, no real progress has been made.

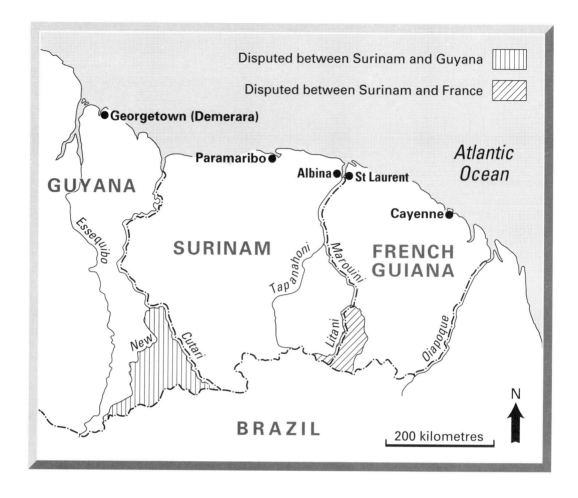

Status

Surinam is in many ways a classic Third World one-product economy with a legacy of incompetent military rule. Furthermore, the ethnic and religious mix is such that rivalries and conflict are always a possibility. The country is a key supplier to the West, particularly the United States of bauxite, but is poverty-stricken. Any interruption to supplies is likely to be of much more than local concern. Therefore, Surinam is likely to remain at least a regional flashpoint for the foreseeable future.

References

Child, J. (1985), *Geopolitics and Conflict in South America*, Praeger/Hoover Institution Press, Stanford.

Day, A.J. (ed.) (1984), *Border and Territorial Disputes*, Longman, London.

The Economist Atlas (1989), Economist Books, Hutchinson, London.

68 Tacna

Description

Tacna is a town and region in the coastal area of southern Peru, near the present Chilean border, and it is the focus of the Bolivia–Chile–Peru territorial dispute involving, in particular, Bolivia's loss of access to the sea. The whole area is part of the Atacama Desert, one of the driest regions on earth, comprising the western slopes of the Andes and a narrow sandy coastal plain, dissected by incised valleys. The region is economically significant for nitrate production.

History and importance

An agreement was reached between Chile and Bolivia in 1866 for a boundary along latitude 24°S and for profits from the nitrate and guano industries, located between latitudes 23° and 25°S. to be shared. However, 85 per cent of the population of Antofagasta Province, the central province in the region covered by the agreement, was Chilean and this so alarmed Bolivia that in February 1873 it signed a secret Treaty of Alliance with Peru.

After Peru (1875) and Bolivia (1878) seized Chilean nitrate companies, Chile declared war on its two neighbours in 1879 (the 'War of the Pacific'). The combined Peru–Bolivia army was defeated at Tarapaca by the Chileans in November 1879. By the subsequent Treaty of Ancon on 20 October 1883, Peru ceded to Chile unconditionally and in perpetuity, the province of Tarapaca and agreed that the provinces of Tacna and Arica should remain under Chilean administration for ten years before a plebiscite was taken.

In 1884, a truce was agreed between Chile and Bolivia, as a result of which Bolivia lost its nitrate industry and access to the Pacific Ocean. This truce was converted into the Peace Treaty of October 1904, when the absolute and perpetual dominion by Chile over the former Bolivian territory was confirmed and the border was demarcated through 96 points. In addition, the Arica to La Paz railway was to be built at Chilean expense and Bolivia's right to full and free transit through Chilean territory to the Pacific was recognized in perpetuity.

The railway was completed in 1913, but in 1918 Bolivia demanded an outlet via a port in either Tacna or Arica province. In 1920, Bolivia reinforced this demand by calling on the League of Nations to revise the 1904 Treaty. However, the League ruled that only the parties could modify the Treaty, and Chile refused. Furthermore, Chile had failed to provide for the plebiscite in Tacna and Arica provinces and American intervention was called for. As a result, on 3 June 1929, the Treaty of Ancon or the Washington Protocol, awarded Arica to Chile, while Peru recovered Tacna and Bolivia gained nothing. Moreover, a provision in the Protocol, which stated that no territory originally belonging to Peru could be ceded to a third party (that is Bolivia) without Chilean approval, vastly complicated Bolivia's quest for access to the Pacific. One result of this was Bolivia's disastrous attempt to gain access to the Atlantic (1932–5) in the Chaco War with Paraguay. As minor compensation for the enormous cost of the war, Bolivia did gain right of access to the Atlantic via Paraguay and the Parana with the use of Puerto Casado (Paraguay) as a free port.

Between 1962 and 1975, Bolivia broke off diplomatic relations with Chile and, in 1973, the revolutionary regime in Peru made large-scale Soviet arms purchases. The threat of a second Pacific War loomed. This was averted by a number of proposals. In December 1975, in attempting to solve the Pacific-access issue, Chile proposed that Bolivia should cede an equal amount of territory (reportedly the mineral-rich area of Potosi) in return for a corridor, should purchase the Chilean section of the Arica–La Paz railway, should allow Chile full use of the waters of the Lauca River, should demilitarize the corridor and should pay compensation for use of port facilities. This proposal was not well received in Bolivia, but was accepted as a negotiating position.

In November 1976, Chile refused to consider a set of Peruvian proposals, whereby Bolivia would receive a 13.5 km-wide corridor along the Chile-Peru frontier, just over 3 km north of the Arica–La Paz railway. An International Zone would be set up, there would be joint control of the Arica port and Bolivia would be allowed to establish a port under its sole sovereignty in a buffer zone, which could claim territorial waters. The stalemate continued, but in June and July 1982 there was a rumour that Argentina had reached a secret agreement with Peru and possibly Bolivia, that Bolivia's right to access to the Pacific Ocean should be recognized.

The whole issue is of crucial importance to Bolivia and to its relations with Peru and Chile. Access to the Pacific is seen as vital to its economic development, but there is also an ideological aspect. Bolivia has consistently attempted to regain the littoral areas which it lost and this has been a constant irritant in both Peru–Chile relations and also particularly those between Chile and Bolivia. The issue has been used as a unifying national factor by Bolivian politicians, but now there is the additional component of the possible resources of a 200 nml EEZ.

Status

A Chilean–Bolivian commission met in Santiago between 18 and 22 March 1991 to discuss the precise border demarcation. This meeting coincided with Bolivia's 'Day of the Sea' and the president announced that Bolivia was 'today closer than ever' to recovering its outlet to the Pacific.

The issue remains controversial and fundamental to Bolivia and, in contrast to the situation a century ago, Chilean military supremacy has now largely gone. The dispute has been predominantly land-orientated, but the issue of the EEZ has added a further complication and threatened to aggravate Bolivia's feelings of resentment.

The latest move occurred on 19 January 1992, when the president of Peru made a concrete proposal for Bolivian access to the Pacific via the Peruvian port of Ilo. This is located some 1,263 km south of Lima and 462 km from La Paz, and about 125 km from the Chilean border. It was added that the access route would benefit not only Bolivia and Peru, but would 'play a strategic role in Pacific Basin trade, and thus open up opportunities for other nations in the region, such as Brazil and Paraguay'. Access to Ilo will reduce Bolivian dependence on Antofagasta, Arica and Iquiqque (all in Chile), and it is further agreed that Ilo should be a duty-free port. This is a hopeful sign that this long-running dispute may be settled, but the long and painful history of disagreement over territory between the three states must introduce an element of doubt about its definitive resolution.

References

Boundary Bulletin, No. 2 (1991), International Boundaries Research Unit, Durham University.
Child, J. (1985), *Geopolitics and Conflict in South America*, Praeger/Hoover Institution Press, Stanford.
Day, A.J. (ed.) (1984), *Border and Territorial Disputes*, Longman, London.
Downing, D. (1980), *An Atlas of Territorial and Border Disputes*, New English Library, London.
Gamba-Stonehouse, V. (1989), *Strategy in the Southern Oceans*, Pinter, London.
Glassner, M.I. (1970), *Access to the Sea for Developing Land-Locked States*, Martinus Nijhoff, The Hague.
Glassner, M.I. (1983), 'The Transit Problems of Landlocked States: The cases of Bolivia and Paraguay' in E.M. Borgese and N. Ginsburg (eds), *Ocean Yearbook 4*, University of Chicago Press, Chicago, pp. 366–89.
Morris, M.A. (1986), 'E.E.Z. Policy in South America's Southern Cone' in E.M. Borgese and N. Ginsburg (eds), *Ocean Yearbook 6*, University of Chicago Press, Chicago, pp. 417–37.

69 The Tanzam Railway

Description

Extending for 1,680 km, the Tanzam Railway links the copper belt of Zambia with the port of Dar-es-Salaam in Tanzania. Completed by the Chinese in 1975, on the Cape gauge of 3 ft 6 in (1.065 m), the railway is single track. It links with the main line, connecting Shaba Province (Zaïre) and South Africa at Kapiri Mposhi.

History and importance

The development of a railway network in southern Africa was stimulated in 1870-1 by the discovery of the Kimberley diamond pipes in South Africa. Since, in all cases, the railways had to descend from the plateau through the coastal ranges, they needed to take comparatively tight curves and for this, the narrow Cape gauge proved most appropriate as compared to the standard gauge of 4 ft 8.5 in. In November 1885, lines from both Cape Town and Port Elizabeth reached Kimberley via De Ar junction.

In 1886, there was an even greater impetus with the discovery of the Witwatersrand goldfields, and the railway from Kimberley was extended from Johannesburg. Additionally, a line was constructed from Durban to Johannesburg. To the north, spurred on by Rinderpest and the Matabele campaign, railway construction reached a rate of a mile a day and by November 1897, Kimberley was linked with Mafeking and Bulawayo. By 1900, Salisbury (Harare) was linked to Beira, and by 1905 an extension had reached Broken Hill (Kabwe) and in 1910, Katanga (Shaba) Province.

The result is that southern Africa has by far the largest rail network in Africa, with some 32,000 km and an infrastructure involving 12 countries, all on the Cape gauge. Of this network, 21,560 km is in South Africa itself.

With the Unilateral Declaration of Independence, by Rhodesia (Zimbabwe) in 1965, the vulnerability of Zambia's economy and its dependence upon the southern African rail network was emphasized. The oil embargo against Rhodesia effectively isolated Zambia. Although an oil pipeline to Dar-es-Salaam was completed in 1968, the vital export of copper was still vulnerable.

The Benguela Railway, dominated by the mineral trade of Zaïre and Angola, provided, technically, some alternative to the South African rail system, but when Portugal left Angola in 1975, this disappeared. However, by 1975, the Chinese, following Western disinclination to participate, had constructed the Tanzania–Zambia (Tanzam or Tazara) Railway. With problems in both Angola and Rhodesia, this provided what was potentially a key lifeline, but for a variety of reasons, mainly technical, it was rarely fully operational.

Status

Since its completion, the utility of the Tanzam Railway has been hampered by port congestion at Dar-es-Salaam, load carrying capacity and a variety of technical difficulties. Indeed, it provided a drain on Zambia's resources. However, it remains a strategic option, but faces a potential decline in importance, with the opening of the Benguela route and the greater security of South African railways. As a land-locked state, Zambia is vitally dependent upon its rail communications to ports in neighbouring countries and since, besides copper, it is an increasingly important exporter of cobalt, the state of its economy is of more than local interest.

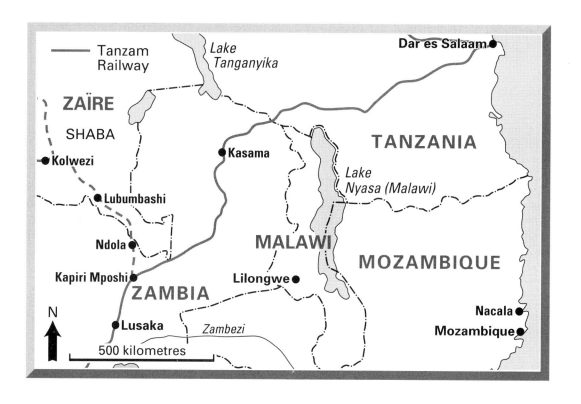

References

Anderson, E.W. (1988), *The Structure and Dynamics of United States Government Policy Making: The Case of Strategic Minerals*, Praeger, New York.

Anderson, E.W. (1988), *Strategic Minerals: The Geopolitical Problem for the United States*, Praeger, New York.

Griffiths, I.L.-L. (1985), *An Atlas of African Affairs*, Methuen. London.

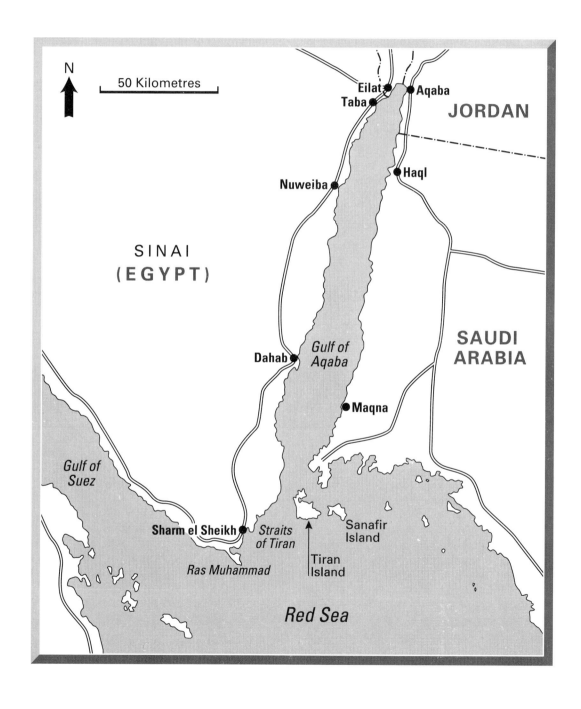

70 The Strait of Tiran

Description

Linking the Red Sea with the Gulf of Aqaba (Eilat) the Strait of Tiran comprises a passage 4 nml wide and 7 nml long, varying in depth between 73 and 83 m. The two main islands in the Strait are Tiran and Sanafir and vessels may only transit through two narrow channels, Enterprise Passage and Grafton Passage, between Sinai and the islands.

The Strait is within the territorial waters of two riparian states: Egypt and Saudi Arabia. Saudi Arabia extended its territorial waters from 3 to 6 nml in 1949, and Egypt followed suit in 1951. Later, in 1958, both extended to 12 nml. However, even before the earlier increase, the territorial waters of both riparians included the Strait in general and the navigable channels in particular. The Gulf of Aqaba itself, varying in breadth from 3 to 14.5 nml is bordered by four coastal states: Egypt, Israel, Jordan and Saudi Arabia.

History and importance

In 1948, soon after the hostilities in Palestine, the Strait was closed to Israeli vessels and cargoes and in 1950, Egypt, with the apparent consent of Saudi Arabia, took possession of the islets of Tiran and Sanafir. After the Suez campaign, in 1957, transit was granted to all vessels of all nationalities and UNEF troops were stationed on the western shore of the Strait at Sharm el Sheikh and Ras Nasrani to guarantee this freedom of passage.

However, in May 1967, Egypt requested the withdrawal of the United Nations forces and again closed the Strait to ships bound for Eilat. In June, after the Six-Day War, Israeli troops occupied the western side of the Strait and the islands. The Strait was re-opened to ships of all flags and the right of innocent passage was guaranteed. In 1979, as part of the implementation of the Camp David Accords, Israel returned the islands to Egypt. They formed part of Zone C, to be supervised by Egyptian police and United Nations forces.

The periodic closure of the Straits to Israeli shipping has been justified by the Arab states on a number of counts. First, they have argued that the Gulf of Aqaba is an internal sea of the Arabs, so that Israel is not considered a legitimate riparian. Despite the multinational status of the Gulf, it has been described as a 'closed Arab Gulf'. If that is accepted, then the principle of free passage does not apply. However, Israel argues that since Egypt, Jordan and Saudi Arabia are undeniably separate states, there is not a single riparian, the Gulf must be multinational and therefore the idea of an internal sea does not apply.

Second, the concept of the historic bay has been promulgated, but the Israeli case is that this is not usually applied in multinational situations. Furthermore, the Gulf has always been used to a considerable extent by foreign, particularly British, ships and this use predates that of the riparians. Therefore, the historic-bay case cannot be sustained.

Third, when a state of belligerency obtains, there is a right to mount a blockade. However, this is not tenable after the end of active hostilities. The fourth point is that, for reasons of history and geography, Israel is not considered a legitimate riparian state.

In fact, the Strait of Tiran appears to fit well into any definition of an international strait. It is a natural maritime passage, including the territorial waters of more than one state, it is used for international navigation, it links two parts of the high seas and it links the high seas to the territorial seas of foreign states. Furthermore, it allows passage from the high seas to the territorial seas of states which are not themselves riparians.

Status

The Gulf of Aqaba, controlled by the Strait of Tiran, is vital as the only maritime outlet for Jordan.

Through Eilat, it also provides Israel with what has proved to be a most important alternative to its Mediterranean coast. During both the Iran–Iraq War and the Gulf Conflict (Operations Desert Shield and Desert Storm), Iraq was largely sustained by Jordan through the port of Aqaba. Given the continuing hostilities between Israel and the Arab world, the Gulf of Aqaba and, in particular, the Strait of Tiran, is likely to remain an important flashpoint.

References

Lapidoth, R. (1975), *Freedom of Navigation With Special Reference to International Waterways in the Middle East*, The Hebrew University, Jerusalem.

Prescott, J.R.V. (1985), *The Maritime Political Boundaries of the World*, Methuen, London.

Times Atlas of the Oceans (1983), Times Books, London.

71 The Gulf of Tongking

Description

The Gulf of Tongking is bounded by Vietnam to the west, the Chinese mainland to the north and Hainan Island (China) to the east. It covers an area of 24,000 nml^2 and has a maximum depth of about 80 m. The Gulf measures 170 nml at its widest and has two outlets: the Hainan Strait between Hainan Island and the Luichow Peninsula, approximately 19 nml in width; and the major passage to the south, 125 nml wide at its narrowest point. In Chinese, the Gulf is known as Beibu and in Vietnamese, Bac Bar.

History and importance

On 26 June 1887, at the Sino-French Convention, the maritime boundary between China and Indo-China (then French, now Vietnam) was apparently delimited. However, it is the outcome of this convention which remains central to the dispute between China and Vietnam. The next event of real significance occurred in April 1973 when North Vietnam was reported to have signed an agreement with Ente Nationale Idrocarburi (ENI), the state petroleum company of Italy, for drilling in the Gulf. Although it was suspected that the agreement had collapsed, the key point was that the outer edge of the proposed ENI Exploration Zone coincided roughly with the median-line boundary between North Vietnam and China.

On 26 December, North Vietnam informed China of its intention to prospect for oil in the Gulf. As the sea boundary had not been delimited, negotiations were proposed. On 18 January 1974, China accepted the offer of negotiations, but stipulated that prospecting should not take place in the area bounded by latitudes 18° and 20°N and longitudes 107° and 108°E and also that no third country should be involved. As a result, on 15 August negotiations opened in Peking (Beijing). However, North Vietnam stood by the 1887 Convention and, by November, with no agreement, the negotiations were suspended.

On 12 May 1977, Vietnam issued a declaration on its territorial sea, EEZ and continental shelf, and on 7 October negotiations on both land and sea boundaries began. In the event, China refused to discuss the land boundary unless Vietnam gave up its claim that the maritime boundary already existed. Stalemate resulted.

On 13 December 1979, Vietnam protested, as a breach of the 1887 Convention, following reports that China had signed contracts with foreign firms to explore for hydrocarbons in the Gulf. Since the outer edge of the proposed Chinese zone approximated to the median line in the Gulf, Vietnam feared that any exploration agreements might eventually result in a *de facto* recognition of Chinese claims.

On 12 November 1982, Vietnam implemented its 1977 declaration on baselines, but the Chinese reaction was to declare the proposed boundaries 'null and void'. Furthermore, it warned that, with regard to 'expansionist designs', the Vietnamese authorities must bear full responsibility for all the serious consequences that might arise therefrom.

The key remains the Sino-French Convention of 1887 in which the relevant section reads: 'The islands which are east of the Paris meridian of 105° 43′E (108° 3′E of Greenwich), that is to say that the north-south line, passing through the eastern part of Taha's Kau or Quan Chou (Tra Co), which forms the boundary, are also allocated to China.'

The Chinese maintain that the maritime boundary issue is unresolved and that the Convention, like other treaties between imperial China and the West, was 'unequal' and is therefore 'null and void'. If the line described were extended to include the whole of the Gulf, it would be more than 130 nml from the coast of Vietnam and only about 30 nml from Hainan, giving approximately two-thirds of the area of the Gulf to Vietnam.

Problems with the Vietnamese claim include the

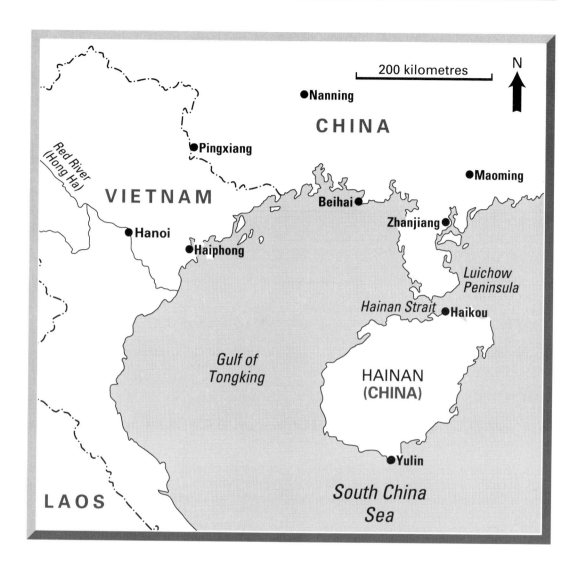

fact that no end points are proposed for the meridian indicated. In the case of boundaries, end points are invariably given. If the line were projected northwards, it would intersect the Chinese coast, thereby effectively awarding waters and sea-bed adjacent to China's coast to Vietnam. If extended south, it would intersect the coast of Vietnam, between Hue and Da Nang. It seems likely that claims would be made east of such a line. Furthermore, if the line were the maritime boundary, Vietnam would not be entitled to any territorial waters off the eastern tip of Quan Chou.

Another more general point is that given the concept of maritime sovereignty at the time, it seems very likely that had the line been intended as a maritime boundary, this fact would have been mentioned in the text. There is in fact nothing distinguishing this line from others used by colonial powers to separate island groups. It was effectively a form of geographical shorthand, seen, for example, in the line suggested in 1939 by the governor general of French Indo-China to settle the dispute over islands between Vietnam and Cambodia in the Gulf of Thailand. But, of course, at the time both states were under the control of the same colonial power and the lack of precision in delimitation matters more to the independent successor states. As an aside, it must be recorded that Vietnam does not accept that particular line, an equidistance line being more favourable.

Status

The Gulf is underlain by thick sediments, an obvious target for oil exploration. So far, no discoveries have been reported, but exploration continues and, given the unsettled nature of the relationships and the problem of defining boundaries, the Gulf of Tongking remain a potential flashpoint.

References

Day, A.D. (ed.) (1984), *Border and Territorial Disputes*, Longman, London.

Dzurek, D.J. (1985), 'Boundary and Resource Disputes in the South China Sea' in E.M. Borgese and N. Ginsburg (eds), *Ocean Yearbook 5*, University of Chicago Press, Chicago, pp. 254–84.

Grinter, L. and Kihl, Y.W. (1987), *East Asian Conflict Zones*, Macmillan, London.

Park, C. (1980), 'Offshore Oil Development in the China Seas: Some Legal and Territorial Issues' in E.M. Borgese and N. Ginsburg (eds), *Ocean Yearbook 2*, University of Chicago Press, Chicago, pp. 302–16.

Prescott, J.R.V. (1985), *The Maritime Political Boundaries of the World*, Methuen, London.

United States Department of State (1983), *Straight Baselines: Vietnam*, Limits in the Seas, No. 99, Bureau of Intelligence and Research, United States Department of State, Washington DC.

72 Transylvania

Description

Transylvania comprises an almost triangular plateau of some 62,000 km^2, with comparatively easy access, north and west, to Hungary, but on the other two sides surrounded by the Carpathian mountains. To the west, the mountains present a largely inpenetrable mass, with only two main passes, but to the south, the Transylvanian Alps are divided into a series of blocks by the deeply incised tributaries of the Danube.

The population of Transylvania is approximately 8 million and includes a high proportion of the Hungarians living in Romania. Romanian statistics give a figure of 1.7 million for the number of Hungarians in Romania, but the statistic promulgated by Hungary is 2 million. The area is predominantly agricultural and economically is only of local importance. However, there are live political issues and Transylvania is a constant factor influencing the relationships between Hungary and Romania.

History and importance

Magyar-speaking settlers began to move into Transylvania in the 10th century, and in 1003 the region was conquered by King Stephen of Hungary. During the 12th and 13th centuries, there was a substantial influx of Germans and the so-called 'Three Nations' (Romanians, Magyars and Germans) enjoyed self-government under the Hungarians and, after 1526, under Ottoman suzerainty. In the 17th century, Transylvania came under the domination of the Austro–Hungarian Empire and in 1848–9, was the scene of violent conflict as the Hungarians attempted to remove Austrian domination. Transylvania was eventually incorporated into the Hungarian part of the Austro–Hungarian Empire in 1867,

In 1916, Romania entered World War I on the side of the Allies with the aim of obtaining Transylvania, and in 1918 it was duly transferred. It remained Romanian until the 'Vienna Award' of August 1940 when the northern half was returned to Hungary. At the same time, Romania lost Bessarabia and Northern Bukovina to the Soviet Union and Southern Dobruja to Bulgaria.

However, in 1941, Romania retook Bessarabia and Bukovina and annexed a large slice of the southern Soviet Union. This situation was reversed after the Soviet victory. At the Paris Peace Treaties between the Allies, Hungary and Romania in 1947, Hungary was forced to return northern Transylvania to Romania.

Until 1956, minorities in Transylvania enjoyed the protection of the Red Army, but during the 1960s, Romania introduced a range of discriminatory measures, including restrictions on the use of the mother tongue. Conditions improved somewhat after 1968 and the Soviet intervention in Czechoslovakia which had led to the fear that the problems of minorities might furnish an excuse for further Soviet activity.

On 24 February 1972, the 1948 Treaty of Friendship, Cooperation and Assistance between Hungary and Romania was renewed for a further 20 years. However, during 1977 and 1978, there were various appeals from the Hungarian minority to the Romanian government for equality and, following the liberation of Eastern Europe and the fall of President Ceauşescu in December 1989, the minorities' issue resurfaced more strongly.

Status

In 1991, riots between Romanians and ethnic Hungarians in Tirgumures in central Transylvania left several dead. The National Salvation Front of Romania has been exploiting, for political reasons, the alleged threat of secession, while the Democratic Association of Hungarians in Romania (RMDSZ) advocates moderation but reports pressures within the minority for a declaration of autonomy. Hungary has long harboured the belief that Transylvania includes part of its natural cultural core and has, as a consequence, been particularly mindful of the problems of the Hungarian minority.

At present, the attentions of Romania are turned more towards the east and Moldova. Should Moldova join with Romania, there would be strong pressures from within for Hungary to attempt to re-absorb at least parts of Transylvania. Therefore, although the boundary between the two countries is not an issue, given the general volatility in Eastern Europe, conflict could well develop.

References

Day, A.J. (ed.) (1984), *Border and Territorial Disputes*, Longman, London.

Downing, D. (1980), *An Atlas of Territorial and Border Disputes*, New English Library, London.

The Independent (1991) 6 November.

73 Trieste

Description

Trieste is situated on the eastern shore, at the head of the Adriatic Sea, facing Venice across the Gulf of Venice. It is located within a narrow coastal extension of the Italian boundary, just north of the Istrian Peninsula. Of all the major ports in Europe, it is Trieste which has suffered most from the many boundary changes in Europe during the 20th century. After World War II, it was the object of complex and protracted negotiations.

History and importance

From the 14th century until 1920, Trieste was under Austrian control and until 1914, it enjoyed a vast hinterland as the principal port of the Austro-Hungarian Empire. On 26 April 1915, at the Treaty of London, Italy agreed to enter World War I on the Allied side, in return for promises of retaining Trieste, South Tyrol and Istria, together with a number of counties and islands. In the event, the Paris Peace Settlement failed to realize all of Italy's high hopes, although it did receive Trentino, Trieste, Istria and the German-speaking part of South Tyrol.

At the Treaty of Rapallo (1920) between Italy and Yugoslavia, Trieste was apportioned to Italy; most of Dalmatia, except the Zara enclave and Lagosta, to Yugoslavia; and Rijeka (Fiume) was designated a free city. However, at the subsequent Italy–Yugoslavia Treaty (27 January 1924) Fiume, already occupied, was given to Italy and Trieste was effectively cut off from its hinterland.

Following World War II, on 15 April 1945, Yugoslavia officially laid claim to Trieste and Istria and on 30 April, Marshal Tito, leader of Yugoslavia, announced that armed forces were in the disputed region. On 1 May, Yugoslav troops occupied Monfalcone and Gorizia, but on 3 May, the New Zealand division captured Trieste. Later that month, the United States and Britain presented notes to Yugoslavia, opposing any unilateral action and stating that the disposal of the disputed territory must form part of the final peace settlement. On 9 June 1945, a temporary military administration of the area was announced and it was laid down that:

(a) part of Istria, including Trieste and the main communications systems, would be under the control of the Supreme Allied Commander (SAC);
(b) Yugoslavian troops, not exceeding 2,000 in number, could occupy within this area a district selected by the SAC but could not have access to other areas;
(c) the SAC would govern through an Allied military government, with a small Yugoslavian observer mission;
(d) Yugoslavian forces must be withdrawn from this area (Zone A) by 12 June;
(e) the Yugoslavian government should return residents to the areas in which they were arrested or from which they had been deported and restitution should be made for damage;
(f) there must be an agreement not to prejudice the final settment of the two zones created:
(A) Allied administration in Trieste and Northern Istria;
(B) Yugoslavian administration in Southern Istria.

On 20 June, the agreement on the demarcation of the boundary between the zones was signed.

In March 1946, a Four-Power (Great Britain, France, the United States and the Soviet Union) Commission of experts was appointed to make the final boundary delimitation, but, by April, had failed to reach unanimous conclusions. Accordingly, on 12 January, the Allied Council of Foreign Ministers, having considered the Commission's report, approved the constitution of a Free Territory of Trieste, with stated boundaries, all territory to the east of which would be ceded to Yugoslavia. Monfalcone and Gorizia were to remain Italian and

the United Nations was to appoint a governor after consultation with Italy and Yugoslavia. This was rejected as unacceptable by both Italy and Yugoslavia.

Thus, on 10 January 1947, the United Nations Security Council took over the administration of the Free Territory. Later, on 10 February, at the Paris Peace Treaty, the Free Territory of Trieste was declared demilitarized and neutral, while Zara and Lagosta were ceded to Yugoslavia. From then until 1975, the saga of Trieste and Zone A continued.

On 20 March 1948, an American–British–French proposal that Trieste should become part of Italy was rejected by Yugoslavia, but later in the year, the Anglo-American Authority issued a decree, bringing the administration of Zone A more into line with that of Italy. On 12 June 1949, the first free elections since 1922 were held in Zone A and of the 60 seats, 40 were gained by Italian parties. One year later, it was announced that Zone B had been completely integrated into the Yugoslavian economy and on 16 April, elections reportedly produced an 89 per cent vote in favour of the Italo-Slovene Popular Front. Later, the Soviet Union called for the withdrawal of Anglo-American troops.

In February 1952, Yugoslavia called for the integration of Zones A and B in one Free Territory under a governor, to be nominated alternately for three years by Italy and Yugoslavia. This was rejected by Italy. On 9 May, a memorandum of understanding from London associated Italy more fully with the administration of Zone A 'in preparation for a final settlement'. Four days later, Yugoslavia stated that the changes amounted to 'an unlawful and unilateral violation' of the Zone's military status. Two days after that, the Yugoslavian military government of Zone B announced a series of measures, designed to link Zone B more closely with Yugoslavia.

On 25 May, municipal elections were held in Trieste and five towns in Zone A, the result being, in every case, an overwhelming majority for the Italian parties. The elections in Zone B, which followed on 7 December, produced, according to the Yugoslavian military government, a 97 per cent vote in favour of the Popular Front. Finally, on 8 October 1953, Britain and the United States jointly stated that they were 'no longer prepared to maintain responsibility for the administration of Zone A' and that they had 'therefore decided to terminate the Allied Military Government, to withdraw their troops and, having in mind the predominantly Italian character of Zone A, to relinquish the administration of that Zone to the Italian government'.

The next day, Yugoslavia protested about what it described as a unilateral violation of the Italian Peace Treaty of 1947 and stated that such an outcome would be unjust because it involved the cession of territory inhabited by Slovenes and it would cut Trieste off from its natural hinterland. In December 1953, Italian and Yugoslavian troops, deployed along their joint border since August, were withdrawn.

The result of all these proposals and counter-proposals was a memorandum of understanding between Italy, Yugoslavia, Britain and the United States (5 October 1954) which gave control of Zone A, apart from a strip of some 13 km^2, inhabited by Slovenes, to Italy and control of Zone B, together with the strip, to Yugoslavia. Additionally, provisions were made for Trieste to become a free port. The memorandum was implemented on 26 October, but the 1954 memorandum was a *de facto* rather than a *de jure* solution. This point was made by Italy in February 1974 when it objected to the erection of border markers by Yugoslavia.

On 1 October 1975, the basis for agreement was reached. A slight alteration in the frontier near Gorizia involving some 3.2 km^2 was made in Italy's favour, and thus, in all, Italy received 233 km^2, with a population of over 302,000 (275,000 in Trieste); and Yugoslavia was given about 320 km^2, with a population of 73,500. Finally, on 10 November 1975, at the Treaty of Osimo (Ancona), the agreement of 1 October was formalized.

Status

The issue of Trieste and the adjacent areas provided a trial of strength between the superpowers during the early stages of the Cold War. There is now no dispute, but Trieste still lacks a hinterland and Slovenia lacks its most economical access to the sea. When the final demise of Yugoslavia is complete, there may well be scope for cooperation between Italy and one or more of the newly recognized states.

References

Day, A.J. (ed.) (1984), *Border and Territorial Disputes*, Longman, London.

Chambers World Gazetteer (1988), Cambridge University Press, Cambridge.

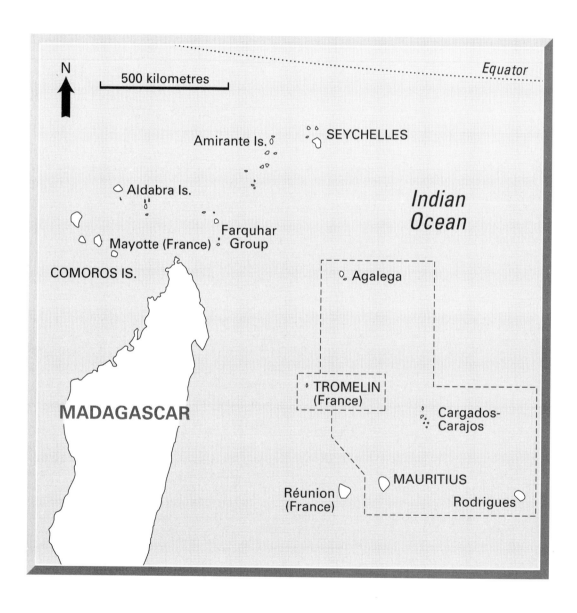

74 Tromelin Island

Description

Located at 15° 52'S, 54° 25'E, Tromelin is 280 nml east of Madagascar and approximately 340 nml north of Mauritius and Réunion. It is tiny, approximately 1 km^2 in area and comprises a volcano, rising directly, some 4,000 m from the ocean floor, topped by a coral plateau. It has no intrinsic economic importance, and access from the sea is difficult. There is no water and no agriculture, but a landing strip is available for light aircraft.

History and importance

Tromelin was discovered by French sailors in 1722 and in 1776 named after a Frenchman who landed on it. However, there is no evidence of France ever having announced occupation, and the island was not in dispute until the 20th century. On 7 May 1954, a permanent French meteorological station was established and this remains the date of effective occupation. However, at a World Meteorological Organization (WMO) Congress, in 1959, Mauritius stated that it considered Tromelin to be part of its own territory, a suggestion refuted by France.

One year later, when Madagascar, from which Tromelin had been administered, became independent, it also laid claim to the island. Nevertheless, in 1976, Madagascar waived its claim in favour of Mauritius. Thus, on 2 April 1976, Mauritius officially claimed Tromelin. Basing its case on an interpretation of the wording of the Treaty of Paris (30 May 1814) under which Britain restored certain Indian Ocean islands taken from France in 1810: 'except the Isle of France (Mauritius) and its dependencies, especially Rodrigues and the Seychelles'. The key word is 'especially', which was defined by Mauritius in its claim as 'in particular, among others or notably'. The interpretation is that, in addition to Rodrigues and the Seychelles, there were other minor dependencies (including Tromelin) which remained British and therefore on independence on 12 March 1968 came under Mauritius.

On 17 December 1976, the French government rejected the claim of Mauritius. France interpreted 'especially' in the text of the Treaty to be in equivalent to 'namely' and that therefore Tromelin was French from 1814. In the French text, the term used was 'nommement' and since French was held to be the language of diplomacy at that time, that text was said to be superior. Indeed, on 2 April 1973, Britain stated that there never had been or was any Franco–British dispute over Tromelin. However, on 20 June 1980, Mauritius announced that its constitution had been amended to add Tromelin in a list of dependencies; the island was included in its sea-bed claim.

Status

Tromelin is still occupied by France and the only potential issue of importance is the effect of the island on sea-bed EEZ claims. If resources are found in the area, the debate may be re-opened. However, conflict is most unlikely.

References

Cottrell, A.J., and Hahn, W.F. (1978), *Naval Race or Arms Control in the Indian Ocean*, Agenda Paper No. 8, National Strategy Information Center, Inc., New York.

Day, A.J. (ed.) (1984), *Border and Territorial Disputes*, Longman, London.

Prescott, J.R.V. (1985), *The Maritime Political Boundaries of the World*, Methuen, London.

75 The Tunbs Islands

Description

The Greater and Lesser Tunb Islands lie some 8 nml apart and are both significantly smaller than the neighbouring Abu Musa. The nearest Iranian island, Qeshm, is only 15 nml away and both Tunbs islands are on the Iranian side of the median line.

History and importance

Even more than Abu Musa, the Tunbs are important for their strategic position in relation to the Strait of Hormuz. They lie in the centre of the main tanker routes to and from the Persian/Arabian Gulf and, being firmly under Iranian control, pose a potential threat to western shipping. Other aspects of significance are considered in the entry on Abu Musa (Map 1).

Status

There seems little reason to believe that any change of status will occur, at least in the medium term. Were there no obvious Iranian control, their position still places them firmly within Iranian jurisdiction. Even with development of the trans-peninsula pipelines, the Strait of Hormuz will remain of key significance as an oil life-line to the West and, particularly, Japan. Therefore, the Tunbs Islands will retain a high profile as a result of their strategic location. Furthermore, as further exploration for offshore oil and natural gas continues, their position will strengthen the claims of Iran.

References

Blake, G.H. and Drysdale, A. (1985), *The Middle East and North Africa: A Political Geography*, Oxford University Press, Oxford.
Day, A.J. (ed.) (1984), *Border and Territorial Disputes*, Longman, London.
Hiro, D. (1985), *Iran Under the Ayatollahs*, Routledge & Kegan Paul, London.
Mojtahed-Zadeh, P. (1990), 'Iran's role in the Strait of Hormuz, 1970-1990' in N. Beschorner, St. J.B. Gould and K. McLachlan (eds), *Sovereignty, Territoriality and International Boundaries in South Asia, South West Asia and the Mediterranean Basin*, Proceedings of a seminar held at the School of Oriental and African Studies, University of London, pp. 96-108.
Peterson, J.E. (1985), 'The Islands of Arabia: Their Recent History and Strategic Importance', Arabian Studies **VII**, pp. 23-35.
Prescott, J.R.V. (1985), *The Maritime Political Boundaries of the World*, Methuen, London.
Rais, R.B. (1986), *The Indian Ocean and the Superpowers*, Croom Helm, London.
Swearingen, W.D. (1981), 'Sources of Conflict over Oil in the Persian/Arabian Gulf', *The Middle East Journal*, **35**, No. 3, Summer, Middle East Institute, Washington DC, p. 314-30.

76 The Tyrol

Description

The Tyrol is a region of the high Alps between Italy, Austria and Switzerland. It unites the upper valleys of the Adige, Drava and Inns Rivers and reaches the Brenner Pass. The area is some 13,600 km^2, comprising Bolzano (7,400 km^2) and Trento (6,200 km^2). Bolzano has a population of approximately 430,000 and the population of Trento is about 450,000 (1980). However, the inhabitants of the latter are 98 per cent Italian and the continuing dispute concerns Bolzano (South Tyrol).

In the Austrian census of 1910, there were 216,000 German-speakers, 16,500 Italian-speakers and 6,000 Ladins. The Italian census of 1921 showed 223,000 Germans and Ladins and 20,000 Italians. In the census of 1971, one-third claimed Italian as their mother tongue and two-thirds, German and Ladin. The 1981 census revealed that the number of Italian-speakers had declined by 3.9 per cent and that of German-speakers increased by 3.4 per cent. Ladins represented 3.6 per cent in 1971 and had increased to 4.2 per cent in 1981.

History and importance

The Tyrol was unified in the 13th century and in 1861, Italy became a unified state. At the Vienna Peace Treaty of 1866, the Austro–Italian border was established, taking into account the fact that the population of South Tyrol was almost entirely German-speaking. During the late-19th and earlier-20th centuries, there were demands in Italy for a 'natural' border up to the Brenner Pass. Since it had received a large Italian influx in the 18th and 19th centuries, Austria was prepared to cede Trento but not South Tyrol.

When on 26 April 1915, at the Treaty of London, Italy agreed to enter World War I on the Allied side, it was promised the whole of South Tyrol to the Brenner Pass. Accordingly, on 10 September 1919, at the Treaty of St Germain-en-Laye, in Article 27, Italy received South Tyrol to the Brenner Pass. Austria protested strongly and was given assurances, with regard to minority rights.

As a result of the Treaty, the province of Bolzano immediately united with Trento to form Venezia Tridentina; 13,600 km^2 were apportioned to Italy. There followed a period of intensive 'Italianization' and from a pre-war figure of under 10 per cent, by 1939 the population of Italian origin in Bolzano had risen to 30 per cent. In 1937, the area was redivided into Bolzano and Trento.

Following the Hitler–Mussolini Pact of May 1939, agreement was reached that German-speakers should opt once and for all for German nationality and therefore compulsory resettlement or Italian nationality. Of the 263,000 German-speakers, 205,441 opted for Germany, but only some 75,000 moved. After the war, over 200,000 were allowed to re-acquire Italian citizenship.

In 1943, German troops occupied South Tyrol, and no attempt was made to annexe the area, yet on 22 April 1946, a petition was presented by the inhabitants to Austria, calling for a return. Three days later, Austria delivered a memorandum on the return and the Austrian claim to the United Nations. However, on 5 September at the Paris Agreement, German minority rights were guaranteed and this was built into the Italian Peace Treaty of 10 February 1947. On 22 June 1947, Italy amalgamated Bolzano and Trento into Trentino-Atto-Adige.

From 1956, through to the 1970s, there were clashes, particularly with Italian border guards and, as a result of one particular series of exchanges, on 11 July 1961, Italy introduced visa requirements for all visiting Austrians. However, these were lifted on 14 September 1962. Following representations from Austria, on 27 October 1960, the United Nations General Assembly passed a unanimous resolution, calling for peaceful bilateral talks.

In May 1964, a joint committee of experts was

established and in 1969, a new package of offers won the support of the Sudtiroler Volks Partie (SVP). This involved extensive new powers for Bolzano, with 66 out of 100 public-sector jobs being reserved for German-speakers. In the 1988 election, the SVP won 60 per cent of the vote, but the acceptance of job reservations had enforced divisions and resulted in polarization. This has led to the growth of the Fascist Movimento Sociale Italiano (MSI), the only major national party opposed to granting autonomy.

The area has always been important for strategic reasons which involve not only the Brenner Pass, but also Italy's defensible frontiers. However, with the development of the European Community, such fears appear anachronistic. If this factor can now be dismissed, the question of national pride alone remains.

Status

The strong nationalistic tendencies within South Tyrol remain and with the development of regionalism more widely, there could be pressure for fuller autonomy within the European framework. However, this would probably be resisted by Italy and the possibility of further friction remains.

References

Day, A.J. (ed.) (1984), *Border and Territorial Disputes*, Longman, London.

United States Department of State (1966), *Austria – Italy Boundary*, International Boundary Study No. 58 (revised), 6 August, Office of the Geographer, Bureau of Intelligence and Research, Washington DC.

The New York Times (1992), 'Viva Alto Adige!', 14 June.

77 The Wakhan Panhandle

Description

From the village of Eshkasham at its lower or western end, the Wakhan Panhandle extends 300 km to the Vakhjir Pass on the Chinese border in the east. At its eastern end, the Panhandle is divided by a westward salient of Chinese territory which extends to the Vakhjir Pass. Therefore, the extremity of the Panhandle is east of the Pass and its total length is approximately 350 km. At its widest, where it includes the Nicholas Range, the Panhandle is 65 km wide and it narrows to 18 km at its western entrance.

The Wakhan Panhandle lies on the greatest watershed in the world, the mountain range of the Hindu Kush, with peaks of between 5,000 m and 6,500 m. The small population is scattered among a number of isolated villages and the people are herders since no agriculture is possible. Ethnically, they are Tadjhiks. Fayzabad, to the northwest of the Panhandle, has a population of 65,000 and is the focus of routes to the area; the nearest road from Tajikistan crosses the Pamir Mountains 85 km north of the Wakhan boundary.

Travel within the Panhandle is along trails and tracks and movement by wheeled vehicle is virtually impossible. Throughout the winter the region is snowbound, and between May and July travel is seriously hampered by flooding. There are some 20 passes through the Panhandle.

History and importance

The Wakhan Panhandle is a creation of the 19th-century Anglo-Russian struggle for hegemony in south Central Asia. The British, anxious to maintain a unified buffer area (Afghanistan) between northwest India (what is now Kashmir) and the expanding Russian Empire, established the Panhandle as a buffer strip. The western two-thirds of the region were delimited by the Russo-British Agreement of 31 January 1873, and the demarcation was confirmed by Afghanistan and the Soviet Union between 1947 and 1948. This part of the boundary was also covered by the Treaty concerning the Régime of the Soviet-Afghanistan State Frontier (18 January 1958).

The eastern third, from Sari Qul, was delimited by a Russo-British Exchange of Notes, dated 11 March 1895 and confirmed by the Afghanistan-Soviet Union Treaty on Border Demarcation of 16 July 1981. This last Treaty was, of course, signed during the Soviet occupation of Afghanistan, including the Panhandle. Indeed, the Panhandle was considered so important strategically and, in particular, as a conduit for the movement of arms to the Mujahideen, that it was controlled directly from Soviet Military Headquarters at Tashkent in what is now Kazakhstan.

The Treaty brought forth a storm of Chinese protest. On 22 July, the Chinese foreign ministry declared the Treaty to be 'illegal and invalid' as the Soviet Union had no right to conclude a border treaty involving a line with a third country, since the territory immediately to the north, some 20,000 km², had been in dispute between China and the Soviet Union (Russia) for approximately 90 years. However, China stressed it had no outstanding territorial disputes with Afghanistan and supported the Agreement of November 1963, regulating the 70 km China–Afghanistan border at the eastern end of the Panhandle.

Status

Despite the collapse of the Soviet Union, the Wakhan Panhandle retains a strategic and geopolitical significance. It separates the south Central Asiatic states of the former Soviet Union from the Indian subcontinent. With the continuing problem of Kashmir and discord likely to the north, the Panhandle can still be a key buffer.

Its relatively inaccessible terrain, makes it an important route for arms smuggling, as it was in

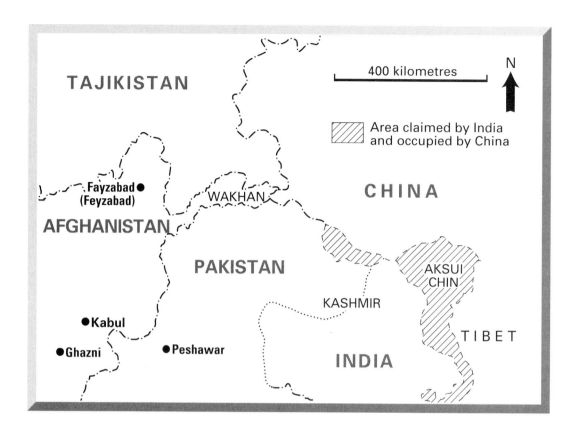

the 1980s during the Afghan Civil War, and in 1990 when support was provided for the Islamic uprising in Xinjiang. In that province, Muslims account for some 60 per cent of the population of 14 million and there is a continuing 'Holy War' to liberate the 'Republic of East Turkestan'.

References

Anderson, E.W. and Dupree, N.H. (eds) (1990), *The Cultural Basis of Afghan Nationalism*, Pinter, London.

Day, A.J. (ed.) (1984), *Border and Territorial Disputes*, Longman, London.

United States Department of State (1983), *Afghanistan – U.S.S.R. Boundary*, International Boundary Study No. 26, revised September 1983, Office of the Geographer, Bureau of Intelligence and Research, Washington DC.

United States Department of State (1974), *China – U.S.S.R. Boundary*, International Boundary Study No. 64, revised 22 January 1974, Office of the Geographer, Bureau of Intelligence and Research, Washington DC.

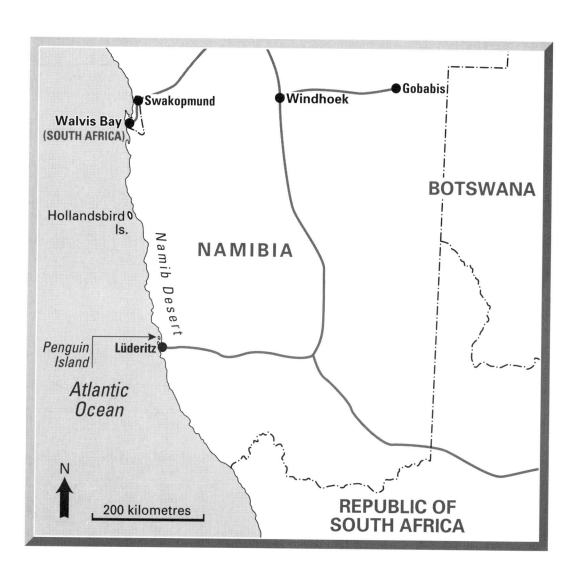

78 Walvis Bay

Description

Walvis Bay is an enclave of some 1,124 km², located midway along the coast of Namibia, 275 km west of Windhoek. Approximately 160 km south of the port of Walvis Bay is Hollandsbird Island, the first of the Penguin Islands, which run in a line, a few miles off the coast, stretching southwards over a distance of some 400 km. The southernmost of these islands is Roastbeef and the others are Ichaboe, Mercury, Long, Seal, Penguin, Halifax Possession, Albatross Rock, Pomona and Plumpudding. These, together with the enclave, are important for fishing, together with sea-bed and EEZ claims.

History and importance

There is reason to believe that the Portuguese explorer, Bartholomew Diaz, reached Walvis in December 1487, but there were no developments until some whaling settlements were established in the 18th century. In 1793, the Dutch occupied Walvis Bay and Halifax Island, but Britain took over in 1795.

On 12 August 1861, the government of Cape Province declared British sovereignty over all 12 Penguin Islands and this claim was restated five years later when a British ship captured Penguin Island. They were proclaimed annexed on 16 July 1866. In 1886 and 1890, the German Colonial Authority recognized the British claim. On 12 March 1878, Walvis Bay was annexed by Britain and later that year, Letters Patent ratified the claim. On 16 August 1884, a German Protectorate was declared over South West Africa and on 1 July 1890, Britain and Germany recognized each other's claims.

After World War I, in 1922, the mandate for the administration of South West Africa, including Walvis Bay, was given by the League of Nations to South Africa. However, on 31 August 1977, under the Walvis Bay Administrative Proclamation, Walvis was detached from South West Africa (Namibia) and re-integrated with Cape Province. This move was condemned by the South West Africa People's Organization (SWAPO) and the United Nations.

Walvis Bay is important as the only deep-water port along the Namibian coast. It deals with the majority of Namibian trade, including the export of uranium from Rossing. It also has significant fishing, canning and fish-meal industries and is ranked fifth among South African ports.

Status

Despite the independence of Namibia in 1990, Walvis Bay is still administered by South Africa, but the SWAPO-based government is committed to completing the decolonization of the country and that concerns, in particular, Walvis Bay. On 21 October 1991, Commonwealth leaders issued a strong call for the re-integration of Walvis Bay and the offshore islands with the mainland, in line with UN Resolution 432 of 1978. Namibia has offered joint administration, but no compromise on the issue of sovereignty.

The future of Walvis Bay is intimately connected with the future of South Africa itself. While under international law South Africa has a legal right to retain Walvis Bay, the likelihood is that, with the coming of majority rule to South Africa, that country will cede Walvis Bay to Namibia. Conflict over the issue seems unlikely.

References

Boundary Bulletin, No. 3 (1992), International Boundaries Research Unit, Durham University, January.

Dale, R. (1982), 'Walvis Bay: A Naval Gateway, an Economic Turnstile, or a Diplomatic Bargaining Chip for the Future of Namibia?' *R.V.S.I.*, **127**, No. 1, March, pp. 31-6.

Day, A.J. (ed.) (1984), *Border and Territorial Disputes*, Longman, London.

Downing, D. (1980), *An Atlas of Territorial and Border Disputes*, New English Library, London.

Griffiths, I.L.-L. (1985), *An Atlas of African Affairs*, Methuen, London.

Munro, D. and Day, A.J. (1990), *A World Record of Major Conflict Areas*, Arnold, London.

Prescott, J.R.V. (1985), *The Maritime Political Boundaries of the World*, Methuen, London.

Prinsloo, D.S. (1977), Walvis Bay and the Penguin Islands: Background and Status, *Foreign Affairs Association Staff Study Report*, No. 8, November, Pretoria.

79 Warbah and Bubiyan Islands

Description

Warbah and Bubiyan are two low-lying, uninhabited islands in an area of mud flats at the northwestern head of the Persian/Arabian Gulf. Warbah is 11 km long and 3 km wide and Bubiyan measures 41 km by 21 km. Bubiyan lies within 1 nml of the Kuwaiti mainland and within 5 nml of Iraq, while the nearest part of Warbah is 2 nml from Kuwait and less than 1 nml from Iraq.

History and importance

In 1899, Kuwait came under British protection, and in 1913 Kuwaiti ownership of Warbah and Bubiyan Islands was recognized in the Anglo–Turkish Agreement. This was, however, never ratified, owing to the onset of World War I. Other key dates are 1932, when Iraq became independent and 1938, when oil was first discovered in Kuwait.

In 1958, the revolution in Iraq heralded the finish of British influence; and with the end of the British Protectorate in 1961, Iraq at once claimed the whole of Kuwait. The claim was based on the fact that Iraq was part of the former Ottoman province of Basra and it was later reinforced by the construction of a commercial port and naval base at Umm Qasr on the Khawr Abd Allah. In fact, Britain had constructed a port there in World War II, but this was dismantled in the post-war period. Under these threats, Kuwait called for British support and troops were duly deployed along the Kuwait–Iraq border, to be replaced later by a joint Arab Force. Two years later, Iraq dropped its claim to the whole of Kuwait but persisted in pressing for Warbah and Bubiyan.

In 1969, there was a border violation near Umm Qasr. This was justified by Iraq, then in dispute with Iran over the Shatt al Arab, as necessary port protection. For a short period in 1972, Iraqi forces entered Kuwaiti territory and there was a further violation in 1973.

With the Iran–Iraq agreement over the Shatt in 1975, Kuwait put pressure on Iraq to remove its forces from the Umm Qasr area. Iraq agreed but suggested that Warbah and Bubiyan should be the subject of a long-term lease agreement. This was rejected by Kuwait and to assert sovereignty and reinforce control, a 4 km bridge was built from the mainland to Bubiyan. Two years later, it was reported that Iraq briefly occupied the islands, but, during the Iran–Iraq War, Kuwait supported Iraq. The Iraqi claim to the islands was revived in 1981 and in 1990 the Gulf Conflict was triggered when the whole of Kuwait was occupied and designated the '19th province' of Iraq. This larger claim presumably subsumed that for the islands, but in early 1991 Iraq was ejected as a result of Operation Desert Storm.

The islands are of no intrinsic economic significance, but they are very important strategically, for two reasons, both related to the tiny length of open coastline (30–40 km, depending upon how it is measured) which Iraq has on the Gulf. First, the islands lie along the southern shore of the Khawr Abd Allah, which provides the only access to the port of Umm Qasr, specifically built to alleviate Iraq's dependence on the vulnerable Shatt al Arab outlet. Second, possession of the islands would allow Iraq to claim a far greater proportion of the maritime and sea-bed resources of the Gulf. Considering that the offshore potential for hydrocarbons is very high, this is a key issue.

Status

The islands remain part of Kuwait, but despite Iraq's defeat in 1991 the dispute is still active. Iraq's restricted access to the Gulf remains an unresolved factor and a particularly sensitive issue. Since the islands would seem to be a natural prolongation of the Iraqi rather than of the Kuwaiti land mass, the issue could well be resolved in Iraq's favour at the International Court of Justice. How-

234 Warbah and Bubiyan Islands

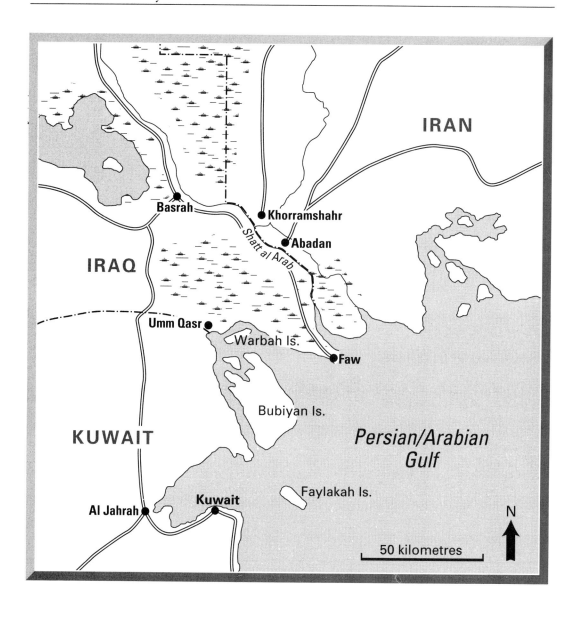

ever, the Court has never been tested and conflict must remain likely over the issue. Given the current problems in Iraq and, in particular, the attitude of Saddam Hussein and his government, this must remain a key world flashpoint.

References

Anderson, E.W. and Rashidian, K. (1991), *Iraq and the Continuing Middle East Crisis*, Pinter, London.

Boundary Bulletin (1992), No. 3, International Boundaries Research Unit, Durham University, January.

Blake, G.H., Dewdney, J. and Mitchell, J. (1987), *The Cambridge Atlas of the Middle East and North Africa*, Cambridge University Press, Cambridge.

Peterson, J.E. (1985), 'The Islands of Arabia: Their Recent History and Strategic Importance', in *Arabian Studies VII*, pp. 23-5.

Prescott, J.R.V. (1985), *The Maritime Political Boundaries of the World*, Methuen, London.

Swearingen, W.D. (1981), 'Sources of Conflict over Oil in the Persian/Arabian Gulf', *The Middle East Journal*, 35, No. 3, Summer, Middle East Institute, Washington DC, pp. 314-30.

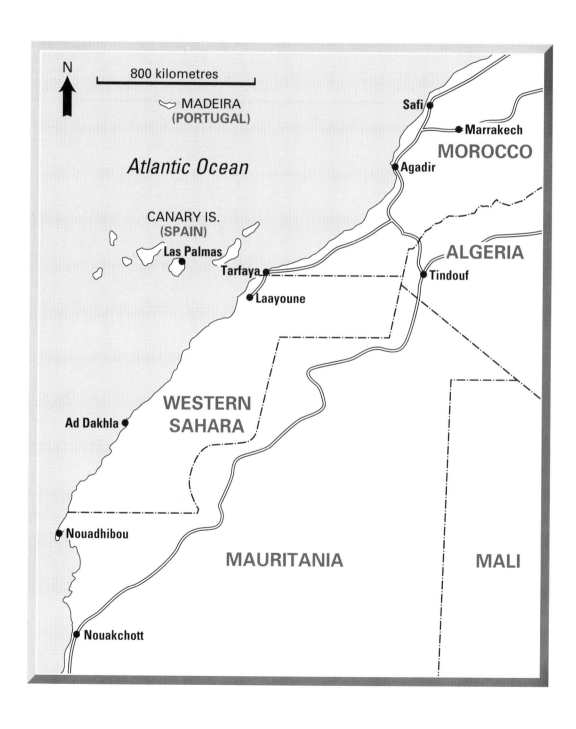

80 Western Sahara

Description

With an area of 270,000 km² and a population of 182,000, Western Sahara is a thinly peopled, mainly desert country. It is located on the west coast of Africa, south of Morocco, with Mauritania on its inland and southern sides. Western Sahara also shares a short length of boundary with Algeria in the extreme northwest and is thus surrounded by three Maghreb countries, all stronger and more developed than itself. Its one notable resource consists of phosphate. The native people are nomadic Sahrawis who were absorbed by the Arabs in the 13th century.

History and importance

Having defeated Morocco in 1860, Spain established, in 1884-6, a protectorate over the southern part of Western Sahara known as Rio de Oro. Morocco itself was then divided into Spanish and French protectorates and the northern part of Western Sahara, Saguia el-Hamra, came under Spanish jurisdiction in 1904. Morocco itself became independent from France in 1956 and rapidly established sovereignty over most of Spanish Morocco by 1958. However, Spain retained the whole of Western Sahara. Morocco's aspirations, nevertheless, extended not only to Western Sahara, but included the whole of Mauritania and the Tindouf region of Algeria.

In 1967, Spain allowed home rule for the Sahrawi people, but Morocco, Algeria and Mauritania had, in the meantime, decided to cooperate to end colonial rule. By the mid-1970s, anti-Spanish forces were joined by the Polisario (Popular Front for the Liberation of Saguia el-Hamra and Rio de Oro) Front, a guerrilla movement, backed initially by Algeria. However, at the same time, the original threefold alignment collapsed, with Morocco and Mauritania both lodging claims for Western Sahara and Algeria supporting independence.

A United Nations' resolution of 14 December 1972 had called for the independence of Spanish Sahara, and Spain announced that in 1975 there would be a referendum to determine the future status of the territory. In the meantime, the United Nations had approved a further resolution on 13 December 1974, sponsored by Morocco and Mauritania, urging that advice should be obtained from the International Court of Justice (ICJ) on the status of Spanish Sahara. On 16 October 1975, the ICJ published its advisory opinion that while 'certain ties of allegiance' existed between Western (Spanish) Sahara and both Morocco and Mauritania ('Mauritanian entity') at the time of colonization by Spain in 1884, these did not support a claim of territorial sovereignty nor affect the principle of self-determination.

Prior to the delivery of the ICJ opinion, the Spanish Council of Ministers announced on 23 May 1975 that it was ready to 'transfer the sovereignty of the Spanish Sahara in the shortest possible period'. Accordingly, the Madrid Agreement was concluded on 14 November 1975. Under this, Spain agreed to withdraw and, despite the ICJ advisory opinion, to hand the territory over to Morocco and Mauritania. Thus, following the departure of Spain (January 1976), Morocco and Mauritania, as secretly agreed between them earlier, partitioned Western Sahara, using the ancient division between Saguia el-Hamra in the north and Rio de Oro in the south. The Polisario Front, backed by Algeria, strongly opposed this act of neo-colonialism and declared an independent Sahrawi Arab Democratic Republic (SADR) and formed a government in exile.

However, following Polsario guerrilla successes and a military coup in Nouakchott (July 1978), Mauritanian troops were withdrawn from Western Sahara (August 1979). Morocco seized the opportunity to take over the southern part of Western Sahara which it renamed Oued Addahab. To support its position, Morocco commenced, in 1980, the building of a defensive wall to seal off Western Sahara from Mauritania. This structure, stretching from wadi Draa in southern Morocco to immediately north of Nouadhibou at Western Sahara's southern boundary, was completed in 1987. Throughout the entire period of its construction, conflict with the Polisario Front continued and a number of mediation attempts by the Organization of African Unity (OAU) and the United Nations failed. Basically, Morocco refused to negotiate with the Polisario, and the Polisario insisted on Moroccan withdrawal as a pre-condition for negotiations. However, in August 1988, both sides accepted a United Nations-sponsored cease-fire formula under which a referendum to determine the wishes of the people of Western Sahara would be held.

The importance of the continuing conflict in Western Sahara is that it has been the focus of confrontation between the states of the Maghreb. As the Maghreb itself becomes more volatile, the issue could be the flashpoint which results in yet more conflict and violence. Furthermore, the situation is exacerbated regionally by the existence of the Spanish enclaves, by the Anglo–Spanish dispute over Gibraltar and by the global significance of Gibraltar Strait as a choke point.

Status

In January 1989, King Hassan of Morocco met with Polisario officials, but no progress was made towards the referendum. The situation was complicated in February 1989 by the formation of the Arab Maghreb Union between Algeria, Morocco, Libya, Tunisia and Mauritania. While not legitimizing the position of Morocco, the new political organization was scarcely likely to support the dismemberment of one of its members to form a new state. Thus, in October 1989, the Polisario Front broke the cease-fire and began a new offensive. However, the Polisario is left with little effective backing.

Despite the United Nations Mission for the Referendum in Western Sahara (MINURSO) with 2,900 members charged with the supervision of the cease-fire and the administration of a referendum to determine either independence or integration into Morocco, Western Sahara is likely to remain a local flashpoint.

References

Boyd, A. (1991), *An Atlas of World Affairs*, ninth edition, Routledge, London.
Day, A.J. (ed.) (1984), *Border and Territorial Disputes*, Longman, London.
Munroe, D. and Day, A.J. (1990), *A World Record of Major Conflict Areas*, Arnold, London.

Index

Abu Masa 1–3, 223
Aegean Sea, The 5–7
Afghanistan 53, 91, 184, 227, 229
Aland Islands, The 8, 9
Albania 57, 61–63, 107, 108
Albanians 10–12, 108
Algeria 52, 169, 177, 237
Alps-Adriatic Region, The (the former Yugoslavia) 10–12
Amnesty International 59
Angola 37, 47
Antarctica 13–15
Aozou Strip, The 16–18
aquifers 74
Arabian Peninsula 23, 141
Arab-Israeli dispute/conflict 71–75, 146, 187, 209, 210
Arafat, Yasser 73
Argentina 13, 35, 36, 67–69, 81, 123
Armenia 135
Armenians 83, 137, 187
Association of South East Asian Nations (ASEAN) 128
Attila Line, The 19–21
Australia 13, 59, 60, 133, 134, 198
Austria 225
Azerbaijan 135–137
Azeris 135

Bab el Mandeb 22–24, 65, 66, 155
Bahrain 85, 91
Baltic Republics, The 25–27
Barents Sea, The 28–30, 97, 103
Basque Country/Euzkadi, The 31–33
bauxite 79, 199–201
Beagle Channel, The 34–36, 123
Begin, Prime Minister 73, 181
Belgium, 13, 57
Benguela Railway, The 37–39, 47
Berlin 40–42
Bessarabia 43–45
Bhutan 122
Bolivia 202–204
Bosnia 11, 12
Botswana 49
Brazil 13, 204
Britain, 1, 9, 13, 19, 23, 25, 35, 49, 51, 53, 65, 67–70, 71–73, 79–81, 89, 101, 113, 121, 133, 143–147, 149, 155, 165, 171, 179, 197, 217–219, 221, 227, 231, 233
British Indian Ocean Territory (BIOT) 53
Bulgaria 13, 107
Burkino Faso 167
Burma 121, 122
Byzantium 135

Cabinda 46, 47
Cape Route 23, 131, 197, 198
Caprivi Strip, The 48, 49
Castro, Fidel 77
Ceuta 50–52
Chad 17
Chagos Archipelago, The (Diego Garcia) 53–55
chemical weapons 111
Chile 13–15, 35. 67, 123, 157, 202–204
China 89, 90, 101, 107, 113, 117, 119, 121, 122, 127, 128, 133, 134, 161, 173–175, 183, 184, 193–195, 198, 205, 211–213, 227
choke points 23, 66, 69, 91, 113, 127, 131, 238
Christians 65, 83, 135, 171, 187, 188
chromium 37
coal mining 191, 198
cobalt 37–39, 205
coke 198
Cold War 29, 41, 57, 77, 117, 219
Commonwealth of Independent States (CIS) 25, 43, 57, 184
communications 21, 23, 51, 89, 135, 205
conservation of the environment 13
continental shelf 5, 69, 165, 211
copper 37, 103, 205
Croatia 10–12
Cuba 37, 77, 155
Curzon Line, The 56, 57, 153, 139
Cyprus 5, 19–21
Czechoslovakia 13, 153

Democratic Association of Hungarians in Romania (RMDSZ) 215
Denmark 13, 95, 165
desertification 167
Djibouti 23, 24, 65, 66, 155
drought 65, 66, 167

East Timor 58–60

239

Egypt 73, 159, 179-181, 197, 198, 209
Ente Nationale Idrocarburi (ENI) 211
Epirus 61-63
Eritrea 24, 64-66, 155
Eritrean People's Liberation Front (EPLF) 65
Estonia 25
Ethiopia 23, 24, 65, 66, 155
Europe 11, 33, 41, 57, 71, 91, 119, 134, 153, 157, 198, 216, 226
European Community (EC) 12, 33, 51, 52, 146, 153, 226
Euzkadi ta Askatasuna (Basque Homeland and Freedom) (ETA) 31, 33
Exclusive Economic Zone (EEZ) 1, 13, 15, 35, 69, 95, 123, 131, 133, 173, 191, 204, 211, 221, 231

fabricated metals 198
Falkland Islands, The (Malvinas) 35, 36, 67-70, 81, 123, 131
famine 65, 66, 167
fertilizer industry 103
fertilizers 198
Finland 9, 97, 103
fish canning and fish-meal industries 231
fisheries 69
fisheries protection zone 191
fishermen 161
fishing 29, 69, 95, 103, 149, 177, 191, 231
fishing rights 173, 191
fish products 103
fish resources/grounds 35, 95, 113, 117, 131
France 9, 13, 17, 33, 35, 47, 53, 65, 67, 71, 83, 89, 97, 129, 133, 134, 155, 157, 161, 171, 179, 187, 193, 199, 217, 219, 221, 237
Franco, General 31-33, 51
free-trade zone 25

Gambia 171
Gaza Strip, The 71-75, 179
Germany 15, 25, 41, 42, 49, 61, 97, 107, 113, 135, 149, 153, 225, 231
Germany, East 13, 41, 153
Germany, West 13, 41
Gibraltar 51, 52, 131, 146, 197, 238
Golan Heights, The 71-75, 188
Gorbachev, President 37, 115, 135
Great Manmade River (GMR) 17
Greece 5, 12, 19-21, 61-63
Greenland 29, 95
Greenpeace 133
groundwater 74
Guantanamo 76-78
Gulf Conflict 181, 210, 233
Gulf Cooperation Council (GCC) 85
Guyana 79-81

Haile Selassi, Emperor 65, 155
Hatay, The 82-84

Hawar Islands, The 85-87
Hercegovina 11, 12
Hindus 101, 199
Hong Kong 25, 88-90, 119, 161, 173
Hormuz, The Strait of 1, 51, 91-93, 223
human rights 59, 75, 102, 111, 137
Hungarian Empire 10-12
Hungary 10-12, 107, 215, 216
Hussein, King of Jordan 73
hydrocarbons 60, 69, 117, 161, 163, 165, 173, 211, 233
hydro-electric power (HEP) 79-81, 103

Iceland 95, 165
India 13, 23, 53, 101, 121, 122, 128, 163, 197, 198, 227
Indonesia 59, 127, 193
industrialization 31, 103
International Court of Justice (ICJ) 5, 17, 35, 67, 85, 95, 165, 237
International Geophysical Year (1957-8) 13
Iran 1-3, 91-93, 109-111, 177, 178, 223, 233
Iran-Iraq War 1, 91, 111, 141, 178, 210, 233
Iraq 1, 73, 91-93, 109-111, 141, 177, 178, 210, 233-235
Iceland 143-147, 165
Irish Republican Army (IRA) 143, 145, 146
iron ore 17, 31
iron oxide-mining 1
irrigation 83, 111, 169, 177
Islamic Fundamentalism 74, 102
Israel 23, 71-75, 179-181, 188, 197, 209, 210
Italy 9, 17, 61, 65, 107, 217-219, 225, 226

Jan Mayen Island 94-96
Japan 13-15, 60, 91, 113-115, 117, 128, 159, 161, 173-175, 183, 193-195, 223
Jordan 23, 71-73, 187, 188, 209, 210

Karelia 97-99, 103
Kashmir 100-102, 121, 163, 227
Kazakhstan 227
Kenya 155
Kola Peninsula, The 29, 103-105, 191
Korea 117
Kosovo 106-108
Kowloon 89, 90
krill 15
Kurdistan 109-111
Kurds 83, 109-111, 177, 187
Kurile Islands, The 113-115
Kuwait 85, 91-93,. 141, 177, 197, 233-235

Latvia 25
League of Nations 9, 61, 71, 83, 187, 231
Lebanon, 73, 83, 187
Liancourt Rocks, The 116, 117
Libya 17, 18, 169, 238
Lithuania 25

Macao 118, 119
Makarios, President 19
Malacca, The Strait of 51, 126-128
Malawi 149
Malaysia 127, 193-195
Malta 95, 197
manganese 15, 37
Mauritania 237, 238
Mauritius 53-55, 221
Magellan Strait, The 123-125
McMahon Line, The 120-122
Mediterranean 5, 52, 91, 128, 179, 187, 197, 198, 210
Mexico 91
metals 103, 198
military base 53-55, 103
mineral resources 5, 13, 29, 35, 199, 203
mineral rights 149
Mitterrand, President 133
Moldova 43, 216
Mongolia 183
Morocco 51, 52, 236, 238
Mouvement Populaire Mahorais (MPM) 129
Movimento Sociale Italiano (MSI) 226
Mubarak, President 181
Mururoa Atoll 132-134
Muslims 11, 65, 71, 101, 109, 129, 135, 171, 187, 188, 198, 229

Nagorno-Karabakh 135-137
Namibia 49, 231
Nasser, President 73, 197
National Front for the Liberation of Angola (FNLA) 37
National Organization of Freedom Fighters (EOKA) 19
National Union for the Independence of Angola (UNITA) 37
natural gas 29, 47, 85, 223
naval base 97, 157, 177
naval-air base 77
Navassa Island 138-139
navigation rights 149
Nehru, Prime Minister 121
Nepal 122
Netherlands 13, 133, 198, 199
Neutral Zones, The 140-142
New Zealand 13, 133, 134, 157, 217
nickel 103
Nigeria 35, 91
nitrate fertiliser 139
nitrate production 203
North Atlantic Treaty Organization (NATO) 5, 29, 57, 95, 103, 153
North Korea 117
Northern Ireland 143-147, 166
Norway 13-15, 29, 95, 191
nuclear-free area 15
nuclear testing 133, 134

nuclear weapons 128
Nyasa, Lake (Malawi) 149-151

Oder-Neisse Line, The 152, 153
Ogaden, The 154-156
oil discovery 163
oil exploitation 69
oil exploration 60, 165, 193, 211-213, 223
oil export 1-3, 178
oilfield 84, 109, 135
oil imports 128
oil products 198
oil production 91, 141, 181
oil reserves 198
oil resources 5, 15, 17, 29, 47, 59, 69, 85, 91, 109, 135, 141, 195
oil security 1-3
oil transport 1-3, 51, 91, 127, 198, 205, 223
Oman 1, 85, 91-93
Operation Desert Shield 198, 210
Operation Desert Storm 23, 53, 74, 93, 111, 198, 210, 233
opium trade 89
Organization of African Unity (OAU) 17, 55, 155, 238
Organization of American States (OAS) 79
Ottoman Empire 11, 12, 109, 177, 215
ozone layer 13

Pacific Rim 91, 161
Pakistan 101, 102, 122, 163
Palestine 71-75, 179, 209
Palestinians 73, 71-75, 187, 188
Palestine Liberation Organization (PLO) 73, 74, 187, 188
Panama Canal, The 157-159
Panhandle 49, 227-229
Papua New Guinea (PNG) 5
Paracel Islands, The 160, 161, 193
Partido Nacionalista Vasco (PNV) 31
Persia 1, 177
Persian/Arabian Gulf 1, 53, 85, 91, 127, 128, 131, 177, 178, 197, 223, 233
Persian Empire 109, 135, 177
Peru 202-204
petroleum 47, 79, 191, 211
Philippines 193-195
phosphate 237
platinum 37
Poland 13-15, 25, 57, 153
pollution 103, 127, 135
Popular Movement for the Liberation of Angola (MPLA) 37, 49
population 1, 9, 11, 19, 25, 31, 41, 43, 51, 53, 59, 61, 65, 67, 71, 79, 83, 89, 97, 101, 103, 107, 109, 113, 119, 129, 133, 135, 139, 143, 155, 161, 167, 171, 179, 183, 187, 191, 199, 203, 215, 219, 225, 227, 229, 237
ports 25, 37, 51, 65, 67, 90, 93, 103, 177, 178, 184, 191, 203, 204, 205, 231

Portugal 47, 53, 59, 119, 127, 205
Protestants 143, 199

Qatar 85, 91

railways 37–39, 47, 103, 157, 159, 167, 184, 203, 204, 205–207
Rann of Kutch, The 162, 163
Red Sea 23, 65, 91, 179, 197, 198, 209
refugees 53, 57, 63, 66, 73, 77, 111, 153, 155, 167, 187
resource geopolitics 15
Revolutionary Front for Independence (East Timor) (FRETELIN) 59
Rockall 164–166
Romania 13, 43, 215, 216
Roman Catholics 143, 199
Ryukyu Islands, The 173–175
Russia 9, 25, 29, 43, 57, 71, 95, 97, 103, 113–115, 128, 134, 135–137, 161, 183, 191
Russian Empire 9, 43, 183, 227

Sadat, President 73, 181
Sahel, The 167–169
Sahrawi Arab Democratic Republic (SADR) 237
Saudi Arabia 1, 85, 91–93, 141, 179, 198, 209
sea-floor mining 15
sea lanes 23
Senegal 167, 171
Senegambia 170–172
Senkaku and Ryukyu Islands, The 173–175
Serbia 10–12, 107, 108
Sharjah (UAE) 1–3
Shatt al Arab, The 176–178
sheep farming/pastoral economy 67, 167
shipbuilding 103
ship repairing 53, 103
Sinai 179, 197–181
Sinai Peninsula and Taba, The 71, 73, 179–181
Singapore 127, 161, 195, 197
Sino-Russian (formerly Soviet) Border, The 183–185
Slovenia 10–12
Somalia 24, 155
South Africa 13, 23, 37, 39, 49, 146, 205, 231
South African Defence Forces (SADF) 49
South Korea 15, 117
South Lebanon 73, 187–189
South Lebanon Army (SLA) 188
South Pacific Commission (SPC) 133
South West Africa People's Organization (SWAPO) 37, 231
Soviet Socialist Republic (SSR) 43
Soviet Union 5, 9, 13–15, 23, 25, 29, 37, 41, 43, 57, 65, 66, 77, 91, 93, 97, 103, 107, 109, 113–115, 117, 127, 135–137, 153, 155, 179, 183–185, 191, 198, 215, 217, 227
Spain 31–33, 50–52, 67, 77, 79, 237–238
Special Air Services (SAS) 146, 171
Spitzbergen (Svalbard) 29, 190–192

Spratly Islands, The 193–195
strategic minerals 15, 37
Sudan 17, 23, 65, 66
Sudtiroler Volks Partie (SVP) 226
Suez Canal, The 23, 51, 73, 131, 157, 159, 179, 196–198
sugar cane 79
Suez Mediterranean pipeline (SUMED) 23, 51, 197, 198
Supreme Allied Commander (SAC) 217
Surinam 79–81, 199–201
Sweden 9, 35, 97, 193
Switzerland 225
Syria 71, 73, 74, 83, 84, 109, 111, 187

Tacna 202–204
Taiwan 15, 69, 193
Tanzam Railway, The 205–207
Tanzania 149
thalweg 177, 178, 199
Tibet 121, 122
Tiran, The Strait of 208–210
Tito, President Josef/Marshal 11, 107, 217
Tongking, The Gulf of 211–213
tourism 139, 181, 195
transit routes/communications 1, 2, 5, 9, 21, 23, 29, 35, 36, 37–39, 47, 49, 51, 89, 91–93, 95, 103, 113–115, 123, 127, 128, 131, 135, 139, 157–159, 161, 167, 177, 178, 184, 197, 198, 203, 204, 205–207, 209, 223, 227, 233
Transylvania 214–216
Trieste 217–219
Tromelin Island 220, 221
Tunbs Islands, The 1, 222, 223
Turkey 5, 17, 19–21, 83, 84, 91, 109–111, 135, 177
Tyrol, The 217, 224–226

Ulster Defence Association (UDA) 146
Ultra-Large Cargo Carriers (ULCC) 198
Union of Soviet Socialist Republics (USSR) 25, 153
United Arab Emirates (UAE) 1, 85
United Kingdom (UK) 5, 143, 146, 165
United Nations 12, 13, 19, 21, 37, 51, 59, 65, 67, 69, 73, 74, 79, 93, 111, 129, 133, 137, 149, 177, 181, 188, 209, 219, 225, 231, 237, 238
United Nations Conference on the Law of the Sea (UNCLOS) 52, 127
United Nations Emergency Force (UNEF) 73, 179, 209
United Nations High Commission for Refugees (UNHCR) 155
United Nations Interim Force in Lebanon (UNIFIL) 188
United Nations Mission for the Referendum in Western Sahara (MINURSO) 238
United Nations Peace-keeping Force in Cyprus (UNFICYP) 19

United States 13–15, 23, 35, 37, 41, 53, 59, 61, 65, 67, 74, 77, 78, 91, 113, 117, 127, 128, 133, 139, 145, 155, 157, 159, 173, 179, 198, 199–201, 217–219
uranium 17, 103, 231
Uruguay 13

Venezuela 79–81, 91
Venice 217
Very-Large Cargo Carriers (VLCC) 127, 198
Vietnam 161, 193–195, 211–213

wadi 141, 197, 238
Wakhan Panhandle, The 49, 227–229
Walvis Bay 230–232
Warbah and Bubiyan Islands 233–235
Warsaw Pact 11, 29
water resources 21, 74, 101, 167

water supply 135, 189
West Bank, The 71–75
Western Sahara 167, 236–238
World Meterological Organization (WMO) 221
World War I 5, 25, 71, 109, 145, 187, 215, 217, 225, 231, 233
World War II 5, 9, 11, 12, 13, 25, 41, 53, 57, 61, 73, 81, 83, 95, 97, 103, 109, 113, 127, 139, 145, 173, 217, 233

Yemen 23, 24, 155
Yugoslavia 10, 11, 61, 107, 108, 217, 219

Zaïre 37, 47
Zambia 37, 49, 205
Zone of Peace, Freedom and Neutrality (ZOPFAN) 128